T0211590

# Lecture Notes in Business Information Processing                    282

Series Editors

Wil M.P. van der Aalst
  *Eindhoven Technical University, Eindhoven, The Netherlands*
John Mylopoulos
  *University of Trento, Trento, Italy*
Michael Rosemann
  *Queensland University of Technology, Brisbane, QLD, Australia*
Michael J. Shaw
  *University of Illinois, Urbana-Champaign, IL, USA*
Clemens Szyperski
  *Microsoft Research, Redmond, WA, USA*

More information about this series at http://www.springer.com/series/7911

Isabelle Linden · Shaofeng Liu
Christian Colot (Eds.)

# Decision Support Systems VII

## Data, Information and Knowledge Visualization in Decision Support Systems

Third International Conference, ICDSST 2017
Namur, Belgium, May 29–31, 2017
Proceedings

Springer

*Editors*
Isabelle Linden
Department of Business Administration
University of Namur
Namur
Belgium

Christian Colot
Department of Business Administration
University of Namur
Namur
Belgium

Shaofeng Liu
Graduate School of Management
University of Plymouth
Plymouth
UK

ISSN 1865-1348    ISSN 1865-1356 (electronic)
Lecture Notes in Business Information Processing
ISBN 978-3-319-57486-8    ISBN 978-3-319-57487-5 (eBook)
DOI 10.1007/978-3-319-57487-5

Library of Congress Control Number: 2017937493

Printed on acid-free paper

This Springer imprint is published by Springer Nature
The registered company is Springer International Publishing AG
The registered company address is: Gewerbestrasse 11, 6330 Cham, Switzerland

# EURO Working Group on Decision Support Systems

EWG-DSS is a Euro Working Group on Decision Support Systems within EURO, the Association of the European Operational Research Societies. The main purpose of the EWG-DSS is to establish a platform for encouraging state-of-the-art high-quality research and collaboration work within the DSS community. Other aims of the EWG-DSS are to:

- Encourage the exchange of information among practitioners, end-users, and researchers in the area of decision systems
- Enforce the networking among the DSS communities available and facilitate activities that are essential for the start up of international cooperation research and projects
- Facilitate the creation of professional, academic, and industrial opportunities for its members
- Favor the development of innovative models, methods, and tools in the field of decision support and related areas
- Actively promote the interest in decision systems in the scientific community by organizing dedicated workshops, seminars, mini-conferences, and conference, as well as editing special and contributed issues in relevant scientific journals

The EWG-DSS was founded with 24 members, during the EURO Summer Institute on DSS that took place at Madeira, Portugal, in May 1989, organized by two well-known academics of the OR community: Jean-Pierre Brans and José Paixão. The EWG-DSS group has substantially grown along the years. Currently, we have over 300 registered members from around the world.

Through the years, much collaboration among the group members has generated valuable contributions to the DSS field, which resulted in many journal publications. Since its creation, the EWG-DSS has held annual meetings in various European countries, and has taken active part in the EURO Conferences on decision-making-related subjects. Starting in 2015, the EWG-DSS established its own annual conferences, namely, the International Conference on Decision Support System Technology (ICDSST).

The current EWG-DSS Coordination Board comprises six experienced scholars and practitioners in the DSS field: Pascale Zaraté (France), Fátima Dargam (Austria), Shaofeng Liu (UK), Boris Delibašić (Serbia), Isabelle Linden (Belgium), and Jason Papathanasiou (Greece).

# Preface

The proceedings of the seventh edition of the EWG-DSS Decision Support Systems published in the LNBIP series presents a selection of reviewed and revised papers from the Third International Conference on Decision Support System Technology (ICDSST 2017) held in Namur, Belgium, during May 29–31, 2017, with the main topic "Data, Information and Knowledge Visualization in Decision Making." This event was organized by the Euro Working Group on Decision Support Systems (EWG-DSS) in collaboration with the University of Namur.

The EWG-DSS series of International Conference on Decision Support System Technology (ICDSST), starting with ICDSST 2015 in Belgrade, were planned to consolidate the tradition of annual events organized by the EWG-DSS in offering a platform for European and international DSS communities, comprising the academic and industrial sectors, to present state-of-the-art DSS research and developments, to discuss current challenges that surround decision-making processes, to exchange ideas about realistic and innovative solutions, and to co-develop potential business opportunities.

The scientific topic areas of ICDSST 2017 include:

- Qualities of Data Visualization
- Data Visualization and Decision Making/Support
- Relational Data Visualization and Decision-Making/Support
- Innovative Data Visualization, Manipulation, and Interaction
- Social and Network Data Visualization
- Textual Data Visualization
- Qualitative Data Visualization
- Process Visualization
- Spatio-temporal Data Visualization and Management
- Environmental Data Visualization
- Visual Information Extraction
- Visual Information Navigation
- Geographic Information Systems and Decision-Making/Support
- Health Information Visualization
- Visualization Supporting Knowledge Extraction
- Big Data Analytics
- Business Intelligence
- Managerial Dashboard
- Knowledge Acquisition and Management
- Knowledge Extraction and Visualization
- Knowledge Communication Through Visualization
- Visualization and Collaborative Decision Making
- Decision Making in Modern Education

This wide and rich variety of themes allowed us, in the first place, to present a summary of some solutions regarding the implementation of decision-making processes in a large variety of domains, and, in the second place, to highlight their main trends and research evolution. Moreover, this EWG-DSS LNBIP Springer edition has considered contributions selected from a triple-blind paper evaluation method, thereby maintaining its traditional high-quality profile. Each selected paper was reviewed by at least three internationally known experts from the ICDSST 2017 Program Committee and external invited reviewers. Therefore, through its rigorous two-stage-based triple-round review, 13 out of 53 submissions, which correspond to a 24.5% acceptance rate, were selected in order to be considered in this 7th EWG-DSS Springer LNBIP edition.

In this context, the selected papers are representative of the current and relevant DSS research and application advances. The papers are organized in four sections:

(1) Visualization Case Studies. Four case studies are presented in this section: "A Visual Decision Support System for Helping Physicians to Make A Decision on New Drugs," by Jean-Baptiste Lamy, Adrien Ugon, Hélène Berthelot and Madeleine Favre; "A Business Intelligence System for Automatic Traffic Enforcement Cameras," by Mali Sher and Guy Shifrin; "Multicriteria Decision Making for Healthcare Facilities Location with Visualization Based on FITradeoff Method" by Marta Dell'Ovo, Eduarda Asfora Frej, Alessandra Oppio, Stefano Capolongo, Danielle Costa Morais, and Adiel Teixeira de Almeida; and "Business Process Modelling and Visualization to Support E-Government Decision Making: Business/IS Alignment" by Sulaiman Alfadhel, Shaofeng Liu and Festus Oderanti.

(2) Visualization Perspectives. This section gathers propositions for problem-specific visualization, and a study on the use of visualization: "Exploring Visualization for Decision Support in FITradeoff Method and Its Evaluation with Cognitive Neuroscience" by Adiel Teixeira De Almeida and Lucia Reis Peixoto Roselli; "Incorporating Uncertainty into Decision-Making: An Information Visualization Approach" by Mohammad Daradkeh and Bilal Abul Huda; "Process Analytics Through Event Databases: Potentials for Visualizations and Process Mining" by Pavlos Delias and Ioannis Kazanidis; and "Value of Visual Analytics to South African Businesses" by Wisaal Behardien and Mike Hart.

(3) Analytics and Decision. Broadening the scope to analytics concerns, the third section proposes: "Conceiving Hybrid What-If Scenarios Based on Usage Preferences" by Mariana Carvalho and Orlando Belo; "A Semantics Extraction Framework for Decision Support In Context-Specific Social Web Networks" by Manuela Freire, Francisco Antunes, and João Paulo Costa; and "A Method for Energy Management and Cost Assessment of Pumps in Waste Water Treatment Plant" by Dario Torregrossa, Ulrich Leopold, Francesc Hernández-Sancho, Joachim Hansen, Alex Cornelissen, and Georges Schutz.

(4) MCDM. Finally, coming to the fundamentals of decision support, two papers on MCDM conclude our proceedings: "Implementation of an Extended Fuzzy VIKOR Method Based on Triangular and Trapezoidal Fuzzy Linguistic Variables and Alternative Deffuzication Techniques" by Nikolaos Ploskas, Jason Papathanasiou, and Georgios Tsaples; and "Integrating System Dynamics with Exploratory MCDA for Robust Decision-Making" by Georgios Tsaples, Jason Papathanasiou, and Nikolaos Ploskas.

We would like to thank many people who contributed greatly to the success of this LNBIP book. First of all, we would like to thank Springer for giving us the opportunity to guest edit the DSS book, and we especially wish to express our sincere gratitude to Ralf Gerstner and Christine Reiss, who provided us with timely professional guidance and advice during the volume editing process. Secondly, we need to thank all the authors for submitting their state-of-the-art work to be considered for this LNBIP volume. All selected papers are of extremely high quality. It was a hard decision for the guest editors to select the best 13. Thirdly, we wish to express our gratitude to the reviewers, who volunteered to help with the selection and improvement of the papers.

Finally, we believe that this EWG-DSS Springer LNBIP volume has made a high-quality selection of well-balanced and interesting research papers addressing the conference main theme. We hope the readers will enjoy the publication!

July 2017

Isabelle Linden
Shaofeng Liu
Christian Colot

# Organization

## Conference Chairs

Boris Delibašić     University of Belgrade, Serbia
Fátima Dargam     SimTech Simulation Technology, Austria
Isabelle Linden     University of Namur, Belgium
Shaofeng Liu     University of Plymouth, UK
Jason Papathanasiou     University of Macedonia, Greece
Rita Ribeiro     UNINOVA – CA3, Portugal
Pascale Zaraté     IRIT/Toulouse 1 Capitole University, France

## Program Committee

Irène Abi-Zeid     FSA – Laval University, Canada
Abdelkader Adla     University of Oran 1, Algeria
Adiel Teixeira De Almeida     Federal University of Pernambuco, Brazil
Carlos Henggeler Antunes     University of Coimbra, Portugal
Francisco Antunes     INESC Coimbra and Beira Interior University, Portugal
Dragana Bečejski-Vujaklija     Serbian Society for Informatics, Serbia
Marko Bohanec     Jožef Stefan Institute, Slovenia
Ana Paula Cabral     Federal University of Pernambuco, Brazil
Guy Camilleri     Toulouse III University/IRIT, France
Hing Kai Chan     University of Nottingham, Ningbo Campus, UK/China
Christian Colot     University of Namur, Belgium
João Costa     Faculty of Economics University of Coimbra, Portugal
Csaba Csaki     University College Cork, Ireland
Fatima Dargam     SimTech Simulation Technology, Austria
Pavlos Delias     Kavala Institute of Technology, Greece
Boris Delibasic     University of Belgrade, Serbia
Alex Duffy     University of Strathclyde, UK
Sean Eom     Southeast Missouri State University, USA
Jorge Freire De Sousa     University of Porto, Portugal
Gabriela Florescu     National Institute for Research and Development
         in Informatics, Romania
Uchitha Jayawickrama     Staffordshire University, UK
Kathrin Kirchner     Berlin School of Economics and Law, Germany
João Lourenço     Universidade de Lisboa, Portugal
Isabelle Linden     University of Namur, Belgium
Shaofeng Liu     University of Plymouth, UK
Bertrand Mareschal     Université Libre de Bruxelles, Belgium
Nikolaos Matsatsinis     Technical University of Crete, Greece
José María Moreno-Jiménez     Universidad de Zaragoza, Spain

| | |
|---|---|
| Jan Mares | University of Chemical Technology, Czech Republic |
| Ben C.K. Ngan | Pennsylvania State University, USA |
| Festus Oderanti | University of Plymouth, UK |
| Jason Papathanasiou | University of Macedonia, Greece |
| Dobrila Petrovic | Coventry University, UK |
| Francois Pinet | Cemagref, France |
| Nikolaos Ploskas | University of Macedonia, Greece |
| Ana Respício | University of Lisbon, Portugal |
| Rita Ribeiro | UNINOVA – CA3, Portugal |
| Alexander Smirnov | Russian Academy of Sciences, Russia |
| Francesca Toni | Imperial College London |
| Stelios Tsafarakis | Technical University of Crete, Greece |
| Yvonne van der Toorn | Tilburg University, The Netherlands |
| Rudolf Vetschera | University of Vienna, Austria |
| Andy Wong | University of Strathclyde, UK |
| Lai Xu | Bournemouth University, UK |
| Pascale Zarate | IRIT/Toulouse University, France |

## Steering Committee – EWG-DSS Coordination Board

| | |
|---|---|
| Boris Delibašić | University of Belgrade, Serbia |
| Fátima Dargam | SimTech Simulation Technology, Austria |
| Isabelle Linden | University of Namur, Belgium |
| Shaofeng Liu | University of Plymouth, UK |
| Jason Papathanasiou | University of Macedonia, Greece |
| Rita Ribeiro | UNINOVA – CA3, Portugal |
| Pascale Zaraté | IRIT/Toulouse 1 Capitole University, France |

## Local Organizing Team

| | |
|---|---|
| Isabelle Linden | University of Namur, Belgium |
| Bertrand Mareschal | Université Libre de Bruxelles Solvay Brussels School of Economics and Management |
| Christian Colot | University of Namur, Belgium |
| Pierrette Noël | University of Namur, Belgium |

# Sponsors

## Main sponsors

Working Group on Decision Support Systems
(http://ewgdss.wordpress.com/)

Association of European Operational Research Societies
(www.euro-online.org)

## Institutional Sponsors

University of Namur, Belgium
(http://www.unamur.be/)

Fonds de la Recherche Scientifique, Belgium
(http://www1.frs-fnrs.be/index.php)

Graduate School of Management, Faculty of Business,
University of Plymouth, UK
(http://www.plymouth.ac.uk/)

University of Toulouse, France
(http://www.univ-tlse1.fr/)

IRIT Institut de Research en Informatique de
Toulouse, France
(http://www.irit.fr/)

SimTech Simulation Technology, Austria
(http://www.SimTechnology.com)

Faculty of Organisational Sciences, University
of Belgrade, Serbia
(http://www.fon.bg.ac.rs/eng/)

UNINOVA – CA3 – Computational Intelligence Research Group Portugal
(www.uninova.pt/ca3/)

University of Macedonia, Department of Marketing and Operations Management Thessaloniki, Greece
(http://www.uom.gr/index.php?newlang=eng)

**Industrial Sponsors**

**Springer**
(www.springer.com)

**Lumina Decision Systems**
(www.lumina.com)

**ExtendSim Power Tools for Simulation**
(http://www.extendsim.com)

**Paramount Decisions**
(https://paramountdecisions.com/)

**1000 Minds**
(https://www.1000minds.com/)

**Professional Society Sponsors**

**IUFRO**
(http://www.iufro.org/)

# Contents

**Multi-Criteria Decision Making**

# Visualization Case Studies

# A Visual Decision Support System for Helping Physicians to Make A decision on New Drugs

Jean-Baptiste Lamy[1,2,3]([✉]), Adrien Ugon[1,2,3], Hélène Berthelot[1,2,3], and Madeleine Favre[4]

[1] LIMICS, Université Paris 13, Sorbonne Paris Cité, 93017 Bobigny, France
jean-baptiste.lamy@univ-paris13.fr
[2] LIMICS, INSERM UMRS 1142, Paris, France
[3] LIMICS, UPMC Université Paris 6, Sorbonne Universités, Paris, France
[4] Department of Primary Care, Université Paris Descartes,
Société de Formation Thérapeutique du Généraliste (SFTG), Paris, France
http://www.lesfleursdunormal.fr

**Abstract.** When new drugs come onto the market, physicians have to decide whether they will consider the new drug for their future prescriptions. However, there is no absolute "right" decision: it depends on the physician's opinion, practice and patient base. Here, we propose a visual approach for supporting this decision using iconic, interactive and graphical presentation techniques for facilitating the comparison of a new drug with already existent drugs. By comparing the drug properties, the physician is aided in his decision task.

We designed a prototype containing the properties of 4 new drugs and 22 "comparator" drugs. We presented the resulting system to a group of physicians. Preliminary evaluation results showed that this approach allowed physicians to make a decision when they were lacking information about the new drug, and to change their mind if they were overconfident in the new drug.

**Keywords:** Knowledge visualization · Overlapping set visualization · Medical decision support · Drug knowledge

## 1 Introduction

Many drugs are available for major indications, such as pain or infectious diseases. Physicians typically have in their mind a "shortlist" of the drugs they usually consider for a prescription in a given indication, and they will prescribe a drug not belonging to the "shortlist" only if none of them is satisfying for a given patient, *e.g.* due to contraindications. However, new drugs regularly come onto the market. When a new drug is available, physicians have to make a decision: whether they include the new drug in their "shortlist" for the corresponding indication.

This decision is very important, because the prescription of new drugs is associated with a higher risk of serious adverse drug events and hospitalizations

I. Linden et al. (Eds.): ICDSST 2017, LNBIP 282, pp. 3–15, 2017.
DOI: 10.1007/978-3-319-57487-5_1

[14,17], and with a higher cost for health insurances because new drugs are costlier [5,24]. The decision is also very difficult to make, and the physician is under a lot of influence [1], from colleagues, pharmaceutical companies, health insurances and patients. For most new drugs, there is no clear "good" or "bad" choice: the right decision depends on the physician's opinion, practice and experience, and also on his patient base. For example, a new drug associated with a high risk of diarrhea (an adverse effect) can be problematic for a physician who has many young children in his patients, because diarrhea can be life-threatening for babies, on the contrary, for another physician with older patients, it might not be a "deal-breaker" problem.

Today, the major source of information on new drugs is the pharmaceutical company sales representative. However, they are not independent from companies and their information might not be reliable, because new drugs involve huge economic interests for pharmaceutical companies. A review showed that the information from pharmaceutical companies never lead to positive impact on health costs or prescribing quality [21]. Another source of information is the expert opinion, typically found in independent medical journals, but these opinions are not tailored to the patient base of the physician and, as we explained in the previous paragraph, there is no "right" decision that can be taken for granted. Moreover, it is often difficult to assess the independence of experts [3].

In medicine, many clinical decision support systems have been proposed for diagnostic or therapy. These systems typically implement the recommendations of clinical guidelines [7], for instance with a rule-based system. However, in our context, there is no clear "right" decision and thus it nearly impossible to establish rules. Therefore, it is not possible to design a rule-based decision support system producing recommendations such as "you should include this drug in your shortlist" or "you should not", because the physician experience and patient base have an important impact on the decision and they can hardly be quantified and coded in rule conditions.

In this paper, we propose a different approach for decision support, based on drug knowledge visualization. Instead of providing recommendations or expert opinions, our decision support system help physicians to compare the properties of the new drug, such as contraindications or adverse effects, with the properties of the already existent drugs for the same indication (*i.e.* the *comparator* drugs). This comparison is complex, due to the huge number of properties involved, and their associated attributes (*e.g.* how to compare half a dozen drugs according to hundreds of adverse effects and their frequency for each drug?). For facilitating the comparison, we implemented three different visualization techniques, based on icons, interactivity and overlapping set visualization. The paper presents the visual decision support tool and preliminary evaluation results, including a comparison of the physician decisions before consulting the system and after.

The rest of the paper is organized as follows. Section 2 describes the knowledge base. Section 3 describes the visualization techniques we used, and the resulting visual interface. Section 4 gives some preliminary evaluation results. Finally, Sect. 5 discusses the methods and the results, and Sect. 6 concludes.

## 2 Knowledge Base

### 2.1 Design

First, we designed a knowledge base on drug properties, allowing the comparison of these properties between drugs. The design of the knowledge base was inspired by the structure of Summaries of Product Characteristics (SPCs), which are the official reference documents on drugs (one SPC per drug). The SPC lists the clinical properties of a drug, including indications, contraindications, interactions, cautions for use, adverse effects, excipients with known effects. SPC are similar to the drug labels, but more detailed because they are destined to health professionals rather than patients. We considered all clinical properties found in SPC.

In a second time, we organized two focus group sessions involving a total of 17 general practitioners (GPs). GPs were asked to annotate SPCs and other documents, and we analyzed and asked them which drug properties were interesting from their point of view, for deciding whether they would consider a new drug for their future prescriptions. Following the focus group's recommendations, we excluded from the knowledge base three pieces of information that were considered of low interest by GPs: (1) cautions for use (*i.e.* recommendations such as "Reduce the dose for elderly patients"), (2) interactions of the first two levels (out of four), keeping only the "contraindicated" and "unadvised" levels, and (3) adverse effects that are both non-serious and non-frequent. Physicians were not interested by cautions for use and interactions of the first two levels because they do not prevent the prescription of the drug (contrary to contraindications or interactions of the higher levels). We also added economic data (daily costs and repayment rates) that were asked by GPs and absent in SPCs.

The knowledge base was edited using Protégé and formalized as an OWL (Ontology Web Language) ontology [18,19,26]. It belongs to the $\mathcal{SHOIQ}(D)$ family of description logics.

### 2.2 Extraction of Drug Knowledge

Drug knowledge was extracted manually by a pharmacist specialized in drug database (HB), from SPCs. This information was completed with the economic data.

We extracted data for four drugs recently made available in France: Antarene codeine® (ibuprofen+codeine, indicated for moderate-to-severe pain), Ciloxan® (ciprofloxacine, indicated for ear infections), Vitaros® (alprostadil, indicated for erectile dysfunction) and Pylera® (bismuth+metronidazole+tetracycline, indicated for treating *H. pylori* stomach infections). We also extracted data for 22 comparator drugs.

## 3 Visual Decision Support System

### 3.1 Visualization Techniques

We combined three advanced visual techniques for comparing drug properties.

First, **VCM icons** (Visualization of Concept in Medicine) [9–11] are icons for representing the main categories of patient conditions, including symptoms, disorders, age classes,... Due to the high number of patient conditions, it is not possible to create a specific icon for each of them. Therefore, VCM icons are created using an *iconic language*. This language includes a set of primitives (5 colors, 35 shapes and 140 pictograms) and grammatical rules for combining the primitives and creating icons. For representing a patient condition, an icon is made of a basic shape (a circle for physiological conditions or a square for pathological conditions) associated with a color (red for current conditions, brown for past conditions, orange for risk of future conditions) and a white pictogram inside the shape (indicating the organ involved, *e.g.* heart or kidney, or the age class). Shape modifiers can be added to specify the type of disorder, *e.g.* a small bacteria for bacterial infection or an upward/downward arrow for hyper- or hypofunctioning of an organ.

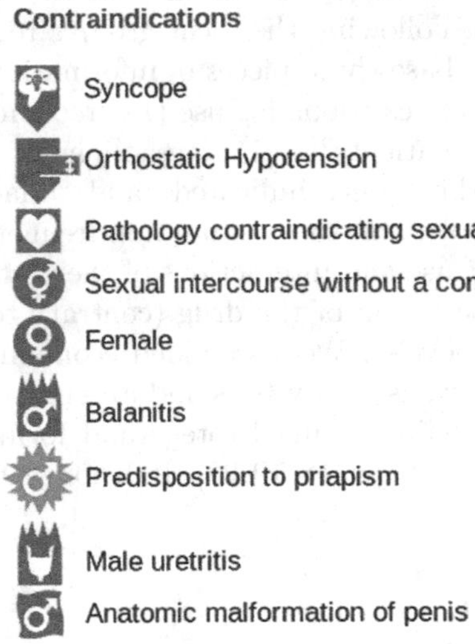

**Contraindications**

Syncope

Orthostatic Hypotension

Pathology contraindicating sexual activity

Sexual intercourse without a condom

Female

Balanitis

Predisposition to priapism

Male uretritis

Anatomic malformation of penis

**Fig. 1.** Example of a list of contraindications with VCM icons.

VCM icons can be used to enrich lists of patient conditions, for example lists of contraindications. The icons can help physicians to quickly identify all contraindications related to a given organ (*e.g.* all cardiac contraindications) or type of disorder (*e.g.* all infections or cancers). Figure 1 shows a synthesis of the contraindications of Vitaros®, with a VCM icon for each.

Second, tables can be used for comparing the numerous clinical drug properties related to security, namely contraindications, interactions and adverse effects. The drugs are displayed in columns and the properties in rows, and the cells contain symbols indicating contraindications or interactions, or the absence of, or small square indicating the frequency of adverse effects (from 0 square, effect is absent, to 5 squares, effect is very frequent). Figure 2 gives an example.

| | Antarene Codeine | Dafalgan Codeine | Izalgi | Lamaline | Ixprim |
|---|---|---|---|---|---|
| *Blood and lymphatic system disorders* | | | | | |
| Haemolytic anaemia | ■ | | | | |
| Agranulocytosis | ■ | | | | |
| *Immune system disorders* | | | | | |
| Drug hypersensitivity | ■ | ■■ | ■■ | ■■ | ■■ |
| Anaphylactic shock | ■ | ■■ | ■■ | ■■ | ■■ |
| Angioedema | ■ | ■■ | ■■ | ■■ | ■■ |
| *Nervous system disorders* | | | | | |
| Meningitis aseptic | ■ | | | | |
| Vertigo | ■■■■■ | ■■■■ | ■■■■ | ■■■■ | ■■■■■ |
| Somnolence | ■■■■ | ■■■■ | ■■■■ | ■■■■ | ■■■■■ |
| *Cardiac disorders* | | | | | |
| Myocardial infarction | ■ | | | | |
| Cardiac failure | ■ | | | | |
| *Respiratory, thoracic and mediastinal disorders* | | | | | |
| Bronchospasm | ■■ | ■■ | ■■ | ■■ | ■■ |
| Asthmatic crisis | ■■■ | | | | |
| Respiratory depression | ■■ | ■■ | ■■ | ■■ | ■■ |
| *Gastrointestinal disorders* | | | | | |
| Pancreatitis | ■ | ■ | | | |
| Haematemesis | ■■■■ | | | | |
| Colitis ulcerative aggravated | ■■■■ | | | | |
| Diarrhoea | ■■■■ | | | | ■■■■ |
| Constipation | ■■■■■ | ■■■■ | ■■■■ | ■■■■ | ■■■■ |
| Dyspepsia | ■■■■ | | | | ■■■■ |
| Flatulence | ■■■■ | | | | ■■■■ |
| Abdominal pain | ■■■■■ | | | | ■■■■ |
| Nausea | ■■■■■ | ■■■■ | ■■■■ | ■■■■ | ■■■■■ |
| Vomiting | ■■■■ | ■■■■ | ■■■■ | ■■■■ | ■■■■ |
| Gastrointestinal perforation | ■■■ | | | | |
| Peptic ulcer | ■■■■ | | | | |
| Crohn's disease aggravated | ■■■■ | | | | |
| Stomatitis ulcerative | ■■■■ | | | | |
| Gastrointestinal haemorrhage | ■■■ | | ■ | ■ | |
| Melaena | ■■■■ | | | | ■■■ |
| *Hepatobiliary disorders* | | | | | |
| Hepatitis | ■ | | | | |
| *Skin and subcutaneous tissue disorders* | | | | | |
| Lyell syndrome | ■ | ■ | ■ | ■ | ■ |
| Stevens-Johnson syndrome | ■ | ■ | ■ | ■ | ■ |
| *Renal and urinary disorders* | | | | | |
| Acute kidney injury | ■ | | | | |
| Show all | | 3 others | 5 others | 4 others | 15 others |

**Fig. 2.** Interactive table showing the adverse effects of Antarene codeine (a new drug for pain) with 4 other drugs with the same indication. All adverse effects of the new drug are shown; for comparators, only the effects shared with the new drug are shown. Below the table, the number of hidden rows is indicated. Serious adverse effects are written in red color. (Color figure online)

**Interactive tables** are an improvement over static tables, in which the rows can be displayed or hidden following the interaction with the user. We propose an interactive table able to display: (1) the properties of the new drug only (the properties of the comparator drugs being displayed only if the new drug has the same property), (2) the properties of the new drug and a user-selected comparator drug, allowing a comparison between two drugs (typically, the new

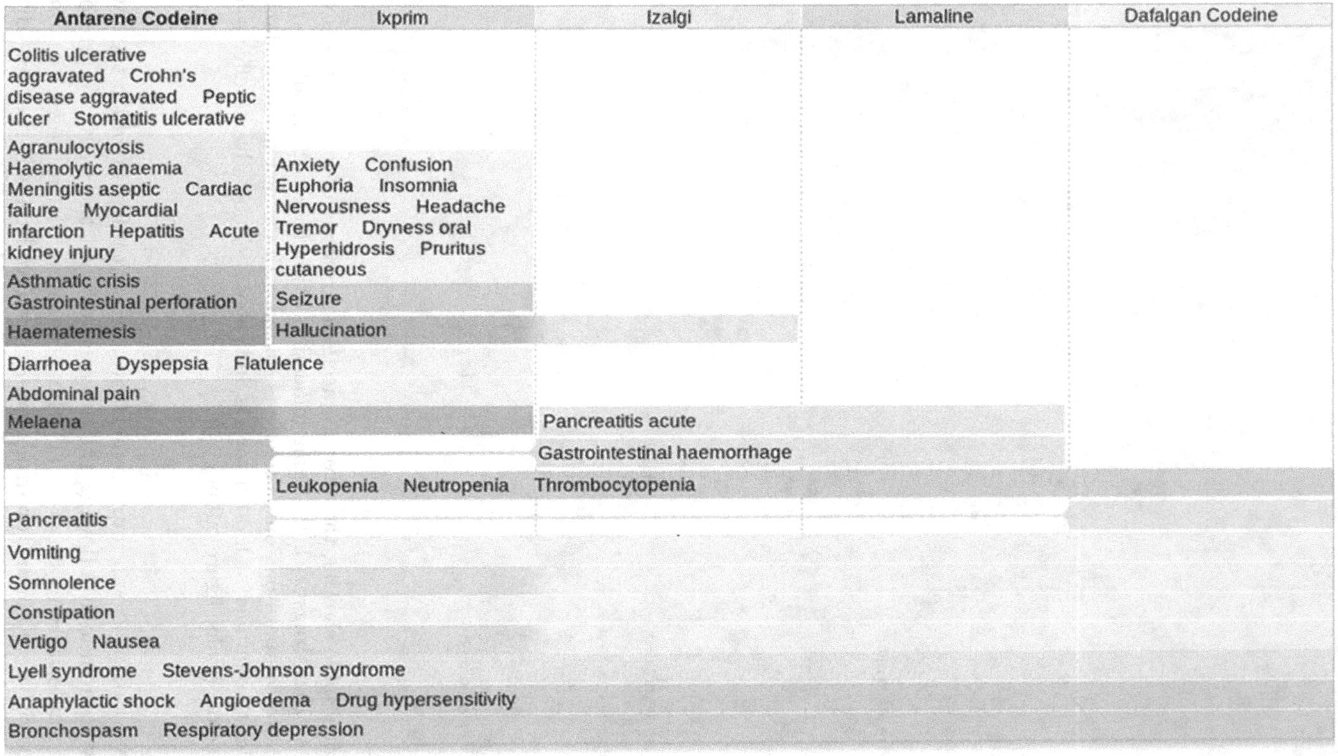

**Fig. 3.** Rainbow boxes comparing the adverse effects of Antarene codeine (a new drug) with 4 comparator drugs (same dataset as Fig. 2, but displaying all effects for all drugs). (Color figure online)

drug and the drug the physician is used to prescribe), (3) the properties shared by the majority of comparator drugs but absent for the new drug (*e.g.* situation in which many comparators are contraindicated but not the new drug), and (4) all properties for all drugs (this often leads to a really huge table).

Third, **rainbow boxes** [8] is an information visualization technique we designed for overlapping set visualization (*i.e.* visualizing several elements and sets made of part of these elements) with drug comparison in mind. For our purpose, the elements are the drugs and the sets are the drug properties. In rainbow boxes, the drugs are displayed in columns and the properties (contraindications, interactions or adverse effects) are displayed as rectangular labeled boxes covering the columns corresponding to the drugs sharing the property. The boxes are stacked vertically, with the largest ones at the bottom. In some situations, "holes" can occur in a box, when the drugs sharing a given property are not in consecutive columns. The drugs are ordered using a heuristic algorithm [8], in order to minimize the number of holes.

For comparing adverse effects in rainbow boxes, we used the color of the boxes to indicate the frequency and the seriousness of the effect. Serious effects are shown in red hue, and non-serious in orange. More frequent effects are shown with more saturated (*i.e.* stronger) colors. In the example of Fig. 3, rainbow boxes allow an easy comparison of the adverse effects of 5 drugs. In the visualization, it is easy to see that Antarene codeine (the new drug on the left) has more adverse effects than other drugs (because there are more boxes in its column), that many of these effects are serious (many red boxes) and that two of these effects are both *frequent* and *serious*, and thus very problematic (Haematemesis and Melaena, easily visible with their strong red color).

In addition to the advanced visual techniques described above, simple tables and bar charts were also used.

## 3.2   Presentation of the Visual Interface

The system was implemented as an HTML website with CSS and JavaScript, the page being generated by Python scripts. Figure 4 shows the general structure of the page for presenting a new drug. In is made of four parts: (1) a title box identifying the new drug, (2) a synthesis listing the properties of the new drug and the name of the similar existing drugs (comparators), (3) a comparative part, comparing the clinical and economic properties of the new drug with the comparators, and (4) a reference part, identifying all drugs and providing links to SPCs. If a new drug has several indications, there is one page per indication, and the list of indications (in the synthesis) includes hyperlinks for navigating through the pages.

The comparative part is the main one. For comparing contraindications, interactions and adverse effects, *i.e.* the three lengthiest and more complex categories of drug properties, it proposes two visualization techniques: interactive table and rainbow boxes. Buttons allow switching from one to the other. For contraindications, VCM icons have also been added two both techniques, as well as the synthesis.

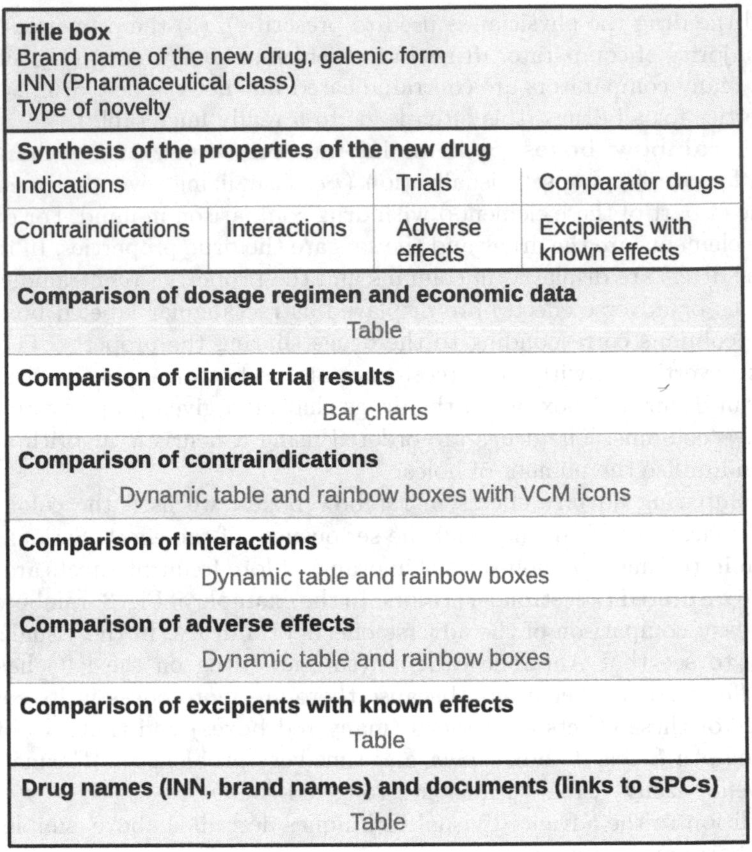

| Title box<br>Brand name of the new drug, galenic form<br>INN (Pharmaceutical class)<br>Type of novelty | | | |
| --- | --- | --- | --- |
| **Synthesis of the properties of the new drug** | | | |
| Indications | | Trials | Comparator drugs |
| Contraindications | Interactions | Adverse effects | Excipients with known effects |
| **Comparison of dosage regimen and economic data**<br>Table | | | |
| **Comparison of clinical trial results**<br>Bar charts | | | |
| **Comparison of contraindications**<br>Dynamic table and rainbow boxes with VCM icons | | | |
| **Comparison of interactions**<br>Dynamic table and rainbow boxes | | | |
| **Comparison of adverse effects**<br>Dynamic table and rainbow boxes | | | |
| **Comparison of excipients with known effects**<br>Table | | | |
| **Drug names (INN, brand names) and documents (links to SPCs)**<br>Table | | | |

**Fig. 4.** General structure of the page presenting a new drug. Categories of drug properties are shown in black, and visualization techniques in blue. (Color figure online)

## 4   Evaluation

The evaluation protocol was difficult to set up for the following reason: there is no possible *gold standard* for the decision of considering the new drug for future prescriptions. The "good" decision rather depends on the GP himself, his conviction, his practice and his patient base. Consequently, during the evaluation, we did not compare the GPs decisions to a gold standard, but we rather evaluated the ability of the visual decision support system to *change* the opinion of the GPs. Thus, we performed a before *vs* after evaluation.

### 4.1   Recruitment

We recruited 22 GPs through an association responsible for the ongoing training of doctors. 12 were men, 10 women, and the mean age was 54.6.

## 4.2   Protocol

The evaluation was carried on a prototype of the website with the 4 new drugs. During the evaluation session, the website was first presented to the GPs in about 20 min, including the various visualization methods described above. GPs were asked to fill a first questionnaire in which they indicated, for each of the 4 new drugs, whether they lacked information about it and whether they were ready to prescribe it in their practice. Then, GPs consulted the decision support system for 45 min. Finally, they completed a second questionnaire with the same questions as in the first questionnaire, and a general discussion was conducted.

## 4.3   Results

88 decisions were collected (22 GPs × 4 new drugs), both *before* and *after* the use of the decision support tool. Each decision was categorized in one of three categories: (1) the GP lacks of information about the new drug (this usually implies that the GP will not prescribe the new drug), (2) the GP has enough information and he is not ready to prescribe the new drug, and (3) the GP has enough information and he is ready to prescribe the new drug.

Table 1 shows the distribution of the 88 decisions among the 3 categories, before and after the use of the decision support tool. In 39 cases, the GP lacked information before; after, in about half of these cases, he decided to retain the new drug for future prescriptions, and he decided not to retain it in the other half. In only 1 case, the GP still lacked information after consulting the decision support tool. In 35 cases, the GP was ready to prescribe the new drug before; in about a third of these cases, he changed his mind after, and decided not to consider the new drug for future prescriptions. In 14 cases, the GP was not ready to prescribe the new drug before; in all of these cases, the GP stayed on his decision.

Therefore, we observed that, when information was lacking, the decision support tool was sufficient for GPs to make a decision in all cases but one. In addition, in some situation, GPs discovered that they were overconfident about a new drug they were ready to prescribe, and they changed their mind.

**Table 1.** Results of the evaluation.

| Before | | After | | |
|---|---|---|---|---|
| | | Not ready to presc. | Lack of info. | Ready to presc. |
| | | 45 | 1 | 42 |
| Not ready to prescribe the new drug | 14 | **14** | 0 | 0 |
| Lack of information about the new drug | 39 | 20 | **1** | 18 |
| Ready to prescribe the new drug | 35 | 11 | 0 | **24** |

# 5   Discussion

In this paper, we presented a visual decision support system for helping physicians to decide whether they should consider a new drug for their future prescriptions. This approach differs from the usual approach for decision support in medicine: the usual approach consists in providing explicit recommendations to the physicians (guided approach) or raising alarms or reminders when the decision made is not the expected one (criticizing approach). However, in the context of new drugs, these usual approaches would have been difficult to implement since, in many situations, there is no "right" decision. On the contrary the appropriate decision depends on the physician experience and his patient base. Moreover, in the medical domain, the acceptance of traditional clinical decision support systems is often low [15].

In the proposed approach, voluminous drug information is provided to the physicians as a decision aid, using visualization techniques for facilitating their consultation. This approach is promising because, with the advent of "big data", more and more information and knowledge is available. According to distributed cognition [12], visualization can amplify the user cognition. In the presented system, the visualizations allow a global overview and a comparison of the properties of the drugs available for a given indication. These visualizations can help the physician to answer several typical questions involved in the decision of whether to consider a new drug for future prescriptions, such as: what are the contraindications of the new drug? Could the new drug be prescribed when existing drugs cannot (due to contraindications)? What are the most frequent adverse effects of the new drug? Does the new drug have fewer adverse effects than existing ones? fewer *serious* adverse effects? How does the new drug compare with the drug X that the physician usually prescribe? The responses to these questions provide argument in favor or against the new drug, and therefore support the decision-making process. This approach also provides a visual explanation for the responses found, thus it might improve the acceptance by physicians.

F. Nake [16] defines *knowledge* as *information* that can be reused in another context, thus acquiring a pragmatic dimension. In that sense, the drug information visualized in this study is actually *knowledge*: the drug properties are determined during clinical trials on a controlled population of patients, and then these properties are considered for prescribing the drug to another patient outside this population (thus another context). In the literature, knowledge visualization has been proposed for knowledge acquisition, knowledge transfer [4] and for improving communication [2], but rarely for decision support. Medical examples include the presentation of antibiotic spectrum of activity for helping physicians to prescribe antibiotics [22], the precise description of a patient with Sleep Apnea Syndrome [23] and VisualDecisionLinc, a visual analytics approach for clinical decision support in psychiatry [13].

The evaluation showed that almost all GPs lacking information were able to find the missing information using the decision support system, and that some GPs that were ready to prescribe the new drug changed their mind after consulting the system. However, we did not observe any GP that was not ready to prescribe the new drug before, and changed his mind. A possible explanation is that GPs not ready to prescribe might have a good reason for that, for instance they might be already aware of a strong negative property of the drug (*e.g.* a serious adverse effects or a high cost). On the contrary, GPs ready to prescribe might ignore some important issue with the drug and, by consulting the system, they can discover it and then change their mind.

Few studies have focused on the comparison of several drugs. C. Wroe *et al.* [25] proposed DOPAMINE, a spreadsheet-like matrix-based tool, but this approach was mostly aimed toward authoring drug properties. Iordatii *et al.* [6] proposed a matrix-based approach for comparing the contraindications and the adverse effects of a new drug to a single reference drug (thus the comparison was limited to two drugs). Drug Fact Boxes [20] offer some comparative drug information, but target patients rather than GPs and does not provide an exhaustive list of the drug properties. More recently, Informulary proposed a drug fact boxes website (http://drugfactsbox.co), but the comparative information is limited to clinical trial results. On the Internet, Iodine (http://www.iodine.com) is a website that collects drug information from patients, including the efficacy of the drug and the adverse events they encountered. Iodine uses tables to compare similar drugs, but the list of the effects of each drug is displayed in a single row, which is tedious for making comparisons. In addition, the quality of patient-collected data is difficult to assess. To conclude, all the proposed approaches were based on tables, whereas our system also relies on icons and rainbow boxes.

# 6   Conclusion

In this paper, we presented a visual decision support system for helping a physician to decide whether he should consider a new drug for his future prescriptions. This system provides several advanced visual tools (icons, interactive tables and rainbow boxes) for facilitating the comparison of a new drug with the already existent drug for the same indication, on the basis of the various clinical and economic drug properties. During a controlled evaluation, the system allowed GPs making a decision on four new drugs: all but one GPs lacking information about a drug obtained enough information to make a decision, and some GPs that were ready to prescribe the new drug changed their mind.

Future works will focus on including additional new drugs in the system, the automatic extraction of drug knowledge from drug database, and the application of the visualization techniques developed for decision support in other domains.

**Acknowledgments.** This work was funded by the French drug agency (ANSM, *Agence Nationale de Sécurité du Médicament et des produits de santé*) through the VIIIP project (AAP-2012-013).

# References

1. Ballantyne, P.J.: Understanding users in the 'field' of medications. Pharmacy **4**(2), 19 (2016)
2. Bertschi, S., Bresciani, S., Crawford, T., Goebel, R., Kienreich, W., Lindner, M., Sabol, V., Vande Moere, A.: What is knowledge visualisation? Eight reflections on an evolving discipline. In: Marchese, E.T., Bannisi, E. (eds.) Knowledge Visualisation Currents: From Text To Art To Culture, vol. 13, p. 32. Springer, London (2013)
3. Bindslev, J.B.B., Schroll, J., Gøtzsche, P.C., Lundh, A.: Underreporting of conflicts of interest in clinical practice guidelines: cross sectional study. BMC Med. Ethics **14**, 19 (2013). doi:10.1186/1472-6939-14-19
4. Burkhard, R.A.: Towards a framework and a model for knowledge visualization: synergies between information and knowledge visualization. In: Tergan, S.-O., Keller, T. (eds.) Knowledge and Information Visualization. LNCS, vol. 3426, pp. 238–255. Springer, Heidelberg (2005). doi:10.1007/11510154_13
5. Garattini, S., Bertele, V.: Efficacy, safety, and cost of new anticancer drugs. BMJ **325**, 269 (2002)
6. Iordatii, M., Venot, A., Duclos, C.: Design and evaluation of a software for the objective and easy-to-read presentation of new drug properties to physicians. BMC Med. Inf. Decis. Making **15**, 42 (2015)
7. Isern, D., Moreno, A.: Computer-based execution of clinical guidelines: a review. Int. J. Med. Inf. **77**(12), 787–808 (2008)
8. Lamy, J.B., Berthelot, H., Favre, M.: Rainbow boxes: a technique for visualizing overlapping sets and an application to the comparison of drugs properties. In: 20th International Conference Information Visualisation, Lisboa, Portugal, pp. 253–260 (2016)
9. Lamy, J.B., Duclos, C., Bar-Hen, A., Ouvrard, P., Venot, A.: An iconic language for the graphical representation of medical concepts. BMC Med. Inf. Decis. Making **8**, 16 (2008)
10. Lamy, J.B., Soualmia, L.F., Kerdelhué, G., Venot, A., Duclos, C.: Validating the semantics of a medical iconic language using ontological reasoning. J. Biomed. Inf. **46**(1), 56–67 (2013)
11. Lamy, J.B., Venot, A., Bar-Hen, A., Ouvrard, P., Duclos, C.: Design of a graphical and interactive interface for facilitating access to drug contraindications, cautions for use, interactions and adverse effects. BMC Med. Inf. Decis. Making **8**, 21 (2008)
12. Liu, Z., Nersessian, N., Stasko, J.: Distributed cognition as a theoretical framework for information visualization. IEEE Trans. Vis. Comput. Graph. **14**, 1173–1180 (2008)
13. Mane, K.K., Bizon, C., Schmitt, C., Owen, P., Burchett, B., Pietrobon, R., Gersing, K.: VisualDecisionLinc: a visual analytics approach for comparative effectiveness-based clinical decision support in psychiatry. J. Biomed. Inf. **45**(1), 101–106 (2012)
14. Moore, T.J., Cohen, M.R., Furberg, C.D.: Serious adverse drug events reported to the food and drug administration, 1998–2005. Arch. Intern. Med. **167**, 1752–1759 (2007)
15. Moxey, A., Robertson, J., Newby, D., Hains, I., Williamson, M., Pearson, S.A.: Computerized clinical decision support for prescribing: provision does not guarantee uptake. J. Am. Med. Inf. Assoc. **17**(1), 25–33 (2010). doi:10.1197/jamia.M3170

16. Nake, F.: Data, information, and knowledge. In: Liu, K., Clarke, R.J., Andersen, P.B., Stamper, R.K. (eds.) Organizational Semiotics: Evolving a Science of Information Systems, Kluwer, Montréal, Québec, Canada, vol. 94, pp. 41–50. Springer, US (2001)

17. Olson, M.K.: Are novel drugs more risky for patients than less novel drugs? J. Health Econ. **23**, 1135–1158 (2004)

18. Rubin, D., Shah, N., Noy, N.: Biomedical ontologies: a functional perspective. Brief. Bioinform. **1**(9), 75–90 (2008)

19. Schulz, S., Jansen, L.: Formal ontologies in biomedical knowledge representation. Yearb. Med. Inf. **8**, 132–146 (2013)

20. Schwartz, L.M., Woloshin, S.: The drug facts box: improving the communication of prescription drug information. Proc. Natl. Acad. Sci. U.S.A. **110**(Suppl 3), 14069–14074 (2013). doi:10.1073/pnas.1214646110

21. Spurling, G.K., Mansfield, P.R., Montgomery, B.D., Lexchin, J., Doust, J., Othman, N., Vitry, A.I.: Information from pharmaceutical companies and the quality, quantity, and cost of physicians' prescribing: a systematic review. PLoS Med. **7**(10), e1000352 (2010)

22. Tsopra, R., Jais, J.P., Venot, A., Duclos, C.: Comparison of two kinds of interface, based on guided navigation or usability principles, for improving the adoption of computerized decision support systems: application to the prescription of antibiotics. J. Am. Med. Inf. Assoc. **21**(e1), e107–e116 (2013)

23. Ugon, A., Philippe, C., Pietrasz, S., Ganascia, J.G., Levy, P.P.: OPTISAS a new method to analyse patients with sleep apnea syndrome. Stud. Health Technol. Inf. **136**, 547–552 (2008)

24. Watkins, C., Harvey, I., Carthy, P., Robinson, E., Brawn, R.: Attitudes and behavior of general practitioners and their prescribing costs: a national cross sectional survey. Qual. Saf. Health Care **12**, 29–34 (2003)

25. Wroe, C., Solomon, W., Rector, A., Rogers, J.: DOPAMINE: a tool for visualizing clinical properties of generic drugs. In: Proceedings of the Fifth Workshop on Intelligent Data Analysis in Medicine and Pharmacology (IDAMAP), pp. 61–65 (2000)

26. Yu, A.C.: Methods in biomedical ontology. J. Biomed. Inf. **39**(3), 252–266 (2006)

# Automatic Traffic Enforcement Camera Operation, Based on a Business Intelligence System

Mali Sher[1(✉)] and Guy Shifrin[2]

[1] Israel Traffic Police, R&D Unit, Beit Dagan, Israel
msher@police.gov.il
[2] Israel Police Technology Department, BI Unit, Jerusalem, Israel
geobisystems@gmail.com

**Abstract.** Since 2012, a new automatic traffic enforcement camera project has been in operation in Israel. Several databases are included in this project, i.e. sensor data, traffic reports, and road accident records. In 2014 a business intelligence system was developed to obtain all the data from the sensors of the new project and to merge them with the existing data to run the project effectively and efficiently. The aim of this paper is to present the process and the configuration of the business intelligence system, and to present the improvements in all measurements. In this paper we demonstrate the importance of a business intelligence system for operating, engineering, researching and managing aspects of a project.

**Keywords:** Business intelligence · Automatic traffic enforcement cameras · Traffic police

## 1 Introduction

The new project of automatic traffic enforcement digital cameras, which detects red-light offenders and speeding drivers, was established in Israel in 2012. Each enforcement site includes electromagnetic loops in the road as sensors of the passing vehicles, and a pole next to the road, which is connected to the sensors. There is an option to install a camera in the pole to record the offences. At the end of 2015 there were around 150 poles and 100 cameras. The data collected from the sensors in each operated site is data per vehicle, which pass on the electronic loops. The data includes time, location, the vehicle's length, a measurement if the vehicle performs an offence, and validation indexes of the data. This database includes around 40 million records per month, therefore it increases rapidly. After two years of gathering these records, it was recognized that a business intelligence (BI) system is crucial for analyzing this database and to merge this new database with other existing databases of traffic reports and road accident records, to optimize the project's benefit for maximal quantity and quality tickets for better deterrence to reduce the number and severity of accidents. To run the project effectively and efficiently, three levels of users have to use this BI system: (1) The control and operation unit that wants to detect timely problems in the flow for

© Springer International Publishing AG 2017
I. Linden et al. (Eds.): ICDSST 2017, LNBIP 282, pp. 16–31, 2017.
DOI: 10.1007/978-3-319-57487-5_2

operating the project smoothly. (2) The research unit responsible for two aspects of strategic and tactical planning: camera mobile planning between the poles in the planning horizon and determining the enforcement's speed level per camera, per period according to the limitation of the resources. (3) The project management's aim is to reduce the bottlenecks and waste in the system. The BI system has to operate throughout all the stages: Detect the sensors' operations, obtain the sensors' data, present the data, convert the data into information, help the different units obtain knowledge from the information, and in the last stage help the workers perform actions to improve the output from the project. The aim of this paper is to present the process and the configuration of the business intelligence system, which was built specifically for the automatic traffic enforcement camera project, and to present the improvements in all measurements as the results. This paper is organized as follows: Sect. 2 is devoted to a literature review. Section 3 describes the stages of building the BI system. Section 4 presents the configuration of the business intelligence system. The results are presented in Sect. 5. The final section summarizes the paper and makes suggestions for further developments.

## 2  Literature Review

In this section, we discuss research in the literature that is covered in this study, including road safety and automatic traffic enforcement cameras (Sect. 2.1), and business intelligence (Sect. 2.2). To the best of our knowledge, this is the first time an integration of the two subjects has been presented.

### 2.1  Road Safety and Automatic Traffic Enforcement Cameras

Approximately 1.25 million people die each year as a result of road accidents worldwide. These road accidents cost approximately 3%–5% of the countries' gross national product [1]. One of the causes of road accidents is traffic offences and speed is one of the main violations that contribute to road accidents, mainly severe ones. For example, for every 1 km/h decrease in the average speed, there is an estimated 4% reduction in the number of accidents [2, 3]. In general, on average, 40–50% of drivers drive faster than the posted speed limit [2, 4–6]. In addition, larger differences in speed between vehicles are related to a higher accident rate [7]. Therefore, an effective road safety strategy needs a balanced approach on three aspects: engineering, education, and enforcement [8].

In this paper, we focus on police enforcement. Traffic police has limited resources, therefore it tries to maximize the benefits from available resources and use effective operation methods. Speed offences can be enforced manually in the form of a police officer with a laser gun who gives a ticket on site, or generates an automatic ticket where the vehicle is spotted as an offender and the owner of the vehicle gets a ticket in the post. Because the traffic police currently has limited resources, they are assigned to tasks that cannot be automated [2, 5, 8].

In 1958, the world's first speed measuring device and the first speed camera used to enforce traffic law was introduced. Since then, many countries have been using speed and red-light enforcement cameras [9]. However, the automatic enforcement camera technology, like any resource, is expensive and cannot be deployed everywhere at all times [5]. Therefore, operational methods should be developed as is described in this research.

## 2.2  Business Intelligence

Organizations have rapidly increasing volume, velocity, and variety of data from many resources and it is necessary to analyze it and make decisions. Business Intelligence (BI) tools help the organization in this aspect [10].

BI systems are a combination of operational data (gathering and storage) and analytical tools for knowledge management at several levels in the organization, i.e. the planners and the decision makers. The BI systems are based on techniques, technologies, methods and applications that analyze and visualize the data to help understand the business performance and make better future decisions to increase value and performance, by time and quality [11, 12]. BI improves the efficiency and effectiveness of the process chains inside the business on one hand, and strengthens the business's relations with customers, suppliers and competitors, and improves strategic and tactical decisions on the other hand [11, 13–15]. The BI system helps the organization in the dynamic business environment with incomplete information, multiple sources of information, and noisy data. It gives a fast response, despite the complexity and several objectives and possible solutions [16]. In addition, the BI system helps to forecast future trends, to implement plans, and to achieve goals in a proactive way [11]. There are several definitions of BI in the literature, as mentioned above. The best definition for our case study is from [10] - A system comprised of both technical and organizational elements that presents its users with historical information for analysis to enable effective decision making and management performance.

BI systems are currently moving from "data-aware" information systems to "process-aware" information systems, because in many organizations the "workflow management" or the "business process management" is the key factor for success in the dynamic and knowledge-intensive world, to improve performance indicators such as cost, quality, time, and flexibility. Therefore, businesses are increasingly interested in improving the quality and efficiency of their processes and the BI systems are changing into a Process-Aware Information System (PAIS) to react to changes in its environment in a quick and flexible way. PAIS helps to manage and execute operational processes involving people and information sources. PAIS uses process mining on the "recording events" or the "event logs" and provides support with the control-flow, based on various forms of analysis (e.g., simulation, monitoring, delta-analyses) and using a constraint-based process modeling approach. There are several options for PAIS. In the workflow management systems, mainly a Person-to-Application (P2A) system is being used, since its primary aim is to make people and applications work in an integrated manner [17–20].

Best practice examples for BI/PAIS systems are mainly from the telecommunications, retail, aviation, finance, healthcare, and automotive industries. The objective is to increase the quality of the service units in these industries: call centers, check-in and production processes [18, 19, 21].

The BI literature has been silent on how BI creates business value and only a few studies have been published on BI for a working team at a macro level [22]. Our paper strives to contribute to these subjects. We will describe in this paper our novel BI system for the automatic traffic enforcement camera project as a Process-Aware Information system (PAIS) for several units working as one team for the project, for business value.

## 3  The Stages of Building the BI System

The new digital automatic traffic enforcement camera project was started in 2012 by the Israel traffic police together with a private company as a PPP (public-private-partnership) project. The private firm that won the contract is responsible for the equipment's installation and the operation aspects (the electromagnetic loops, the poles, the cameras and its illumination, and the relevant software). The traffic police is responsible for the enforcement operation aspect (fixing the enforcement speed level per camera, detecting the offences and producing the traffic tickets, sending the tickets to the vehicle owners, answering letters from the drivers, and prosecuting the ticket in court). A byproduct of the project is the traffic parameters data, which is gathered from the installation of the sensors in the roads, for each of the passing vehicles on the electromagnetic loops. This huge database was gathered by the private company only. In mid 2014, a decision was made that all the data from the sensors collected from all the passing vehicles is important for improving the enforcement by combining it with tickets and road accident data. Therefore, a BI interface was selected and the relevant human and technology resources were allocated to build it on Qlikview, the organization's existing BI platform. There were three stages in the process of building a BI system for this project.

### 3.1  Definition of Needs

The new imported data from the sensors is a unique database of information that will be used for three learning purposes: (1) management– summary, trends, and irregular data represented in tables, graphs, and maps as management and control reports; (2) operations – balance and control of the loads at each working station in the enforcement system (encoders, prosecutions, answering drivers' queries, etc.); (3) researchers and engineering – in-depth investigation of the data over time combined with other BI platforms for road accidents and traffic tickets. Each of the purposes was analyzed and the workers were asked what their needs were. An evaluation of all the needs gave rise to the idea of building several screens that will be suited to all the users, each user with his needs to operate the system: view, drill down, quick output for presentation or to use in other software for further analytic work.

## 3.2    Building the Database According to the Information Definition

The strategic decision was to "pull" all the historical data to date and include all the record fields from the sensors and import them weekly to the BI system. Therefore, we now have all the data for all the vehicles that pass on the electromagnetic loops. Each vehicle's record include these fields: location of the vehicle (site number, lane number, and driving direction), time of the event, for vehicles that cross a junction – the color of the traffic light (red, yellow or green) and the crossing time from the beginning of this color, traveling speed, vehicle length and validation parameters for the speed and length data. This database is added to the other databases at the organization: road accidents and traffic tickets as a whole and especially at the project sites, and the current status of the project operation (operated cameras, level of enforcement per camera). Merging all databases gives us the option to "push" analytic results to the different units in the process. Figure 1 demonstrates the project's UML (Unified Modeling Language) diagram.

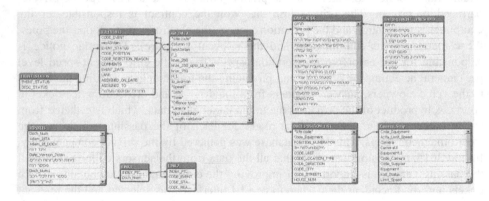

**Fig. 1.** The UML diagram

## 3.3    Analysis of the Data

Several designated screens were developed for all the management-operational-research aspects with comparable GUI (graphical user interface) for the different users. In the next chapter, the configuration of the screens will be demonstrated.

# 4    The Configuration of the BI System

Several screens were developed in this new BI (business intelligence) system based on the process characterization from the previous chapter to address the needs of all the users, taking the sensors' data into information presentation, formatting it into knowledge that will help to perform actions in the automatic traffic enforcement camera project. All the screens are dashboards with interactive graphical displays. The users can have a high-level view or click to choose a specific parameter or different display (graphical, table, or export to Excel).

## 4.1 A Statistical Data Screen

This is the default home page of the BI system, when a user enters the system. It is a table presentation of the statistical data per site as shown in Fig. 2. Each site is a row in the table. The columns are the statistical values that are important for road safety and enforcement policy. The parameters in the columns are: the speed limit, traffic count (number of vehicles crossing the electromagnetic loops per chosen period), total measurement days per chosen period, average speed, the percentage of vehicles driving above the speed limit, standard deviation (Stdev), and the speed on the 85th percentile of the speed distribution. The size of the table is flexible and can be decided by the user: selective periods, select several sites by group (type of camera, type of road, etc.) or a specific site and selective statistical parameters. The table can be sorted according to each of the parameters in the columns. Exceptional values like the percentage of vehicles driving above the speed limit and the standard deviation have been colored according to severity: no color – low importance; yellow – medium concern, and red – high concern.

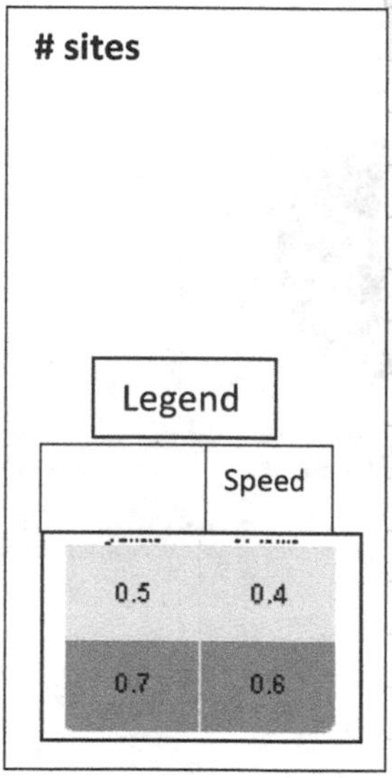

| # sites | Year | 2015 | | | | | | |
| | Speed limit | Traffic volume | # of days | Average speed | % above speed limit | Stdev | Stderr | % 85th |
|---|---|---|---|---|---|---|---|---|
| | 70 | 2,620,368 | 221 | 69 | 0.00% | 12 | 0.00737 | 80.00 |
| | 110 | 4,164,865 | 147 | 95 | 0.00% | 13 | 0.00616 | 105.00 |
| | 100 | 8,892,543 | 239 | 91 | 0.00% | 15 | 0.00506 | 102.00 |
| | 100 | 15,883,502 | 304 | 89 | 15.30% | 16 | 0.00400 | 101.00 |
| | 80 | 2,288,483 | 186 | 74 | 0.00% | 13 | 0.00844 | 85.00 |
| | 70 | 628,835 | 46 | 70 | 0.00% | 16 | 0.01990 | 83.00 |
| Legend | 80 | 1,849,907 | 110 | 76 | 35.70% | 13 | 0.00988 | 87.00 |
| | 90 | 1,729,470 | 107 | 84 | 0.00% | 11 | 0.00856 | 93.00 |
| Speed | 100 | 10,514,687 | 142 | 77 | 10.73% | 27 | 0.00821 | 98.00 |
| | 90 | 11,019,025 | 258 | 87 | 43.08% | 17 | 0.00502 | 99.00 |
| 0.5    0.4 | 80 | 5,426,460 | 357 | 75 | 31.28% | 14 | 0.00590 | 86.00 |
| | 70 | 1,507,166 | 58 | 74 | 64.49% | 14 | 0.01139 | 86.00 |
| 0.7    0.6 | 80 | 1,607,826 | 282 | 77 | 45.31% | 17 | 0.01368 | 90.00 |
| | 80 | 4,014,267 | 351 | 78 | 38.56% | 11 | 0.00556 | 87.00 |

**Fig. 2.** The statistical data screen (Color figure online)

Two models run in the background: (1) summarizing the sensors' data log into the statistical parameters per period and (2) coloring the uncommon and high parameters for visibility.

This screen enables engineering to understand the drivers' behaviors versus the road design. It gives the operation unit recommendations on where to mobilize a camera from an out of order site to a site with the highest percentage of offences. The research unit can evaluate the influence of enforcement on the road safety parameters.

The goal of the enforcement project is to reduce the average and deviation of speed, mainly on the sites that are marked with color.

## 4.2   Traffic Count and the Offences Distribution Screen

This screen graphically displays information and knowledge for the traffic police, of the entire picture of the speed offences in the project as shown in Fig. 3. First is the "big picture", how many offenders there are. Figure 3a demonstrates that in 2015 around 14% of the vehicles that cross the enforcement sites drove above the speed limit, and are therefore offenders. It also showed the total number of vehicles passing the sites, totaling around 350 million. In Israel there are around 3 million registered vehicles [17], thus on average a vehicle passes the project's sites around 117 times per year. Figure 3b zooms into the offender distribution according to level of offence, from low offence with fine and no points, to severe offence with court summons. Figure 3b demonstrates the total offender distribution and Fig. 3c demonstrates a zoom of each site with the offender distribution. This screen gives management a holistic look at the offences in the project and zooms in to specific sites. The goal of the enforcement project is to reduce the number of offenders, mainly the offenders in the top end of the speed distribution.

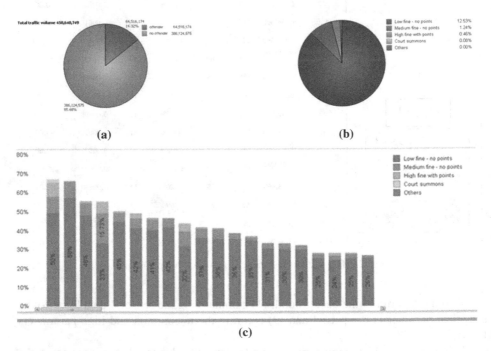

(a)                                      (b)

(c)

**Fig. 3.**   (a) Speed no offenders/offenders distribution. (b) Speed offenders distribution. (c) Speed offenders distribution per site

Two models are running in the background: (1) summarizing and coloring the offences groups per site and per period and (2) summarizing and coloring the sensors' data log into offenders and non-offenders pie.

## 4.3  A Control Panel

A graphical presentation for irregularity in the data is presented in this screen. The model that runs in the background is a summary of all the vehicles in the sensors' data log per month and year. The information is summarized so that it can be seen visually from the graphs. For example, in Fig. 4 the operation users can see the month when the cameras were operated and the number of events per month. This specific camera began to be operated in mid August 2012 to November 2013 and then for brief periods in January–February 2015 and October–December 2015. With this information, the operation unit can follow the operation plan and supervise the private company work. For example, is it according to the plan that there are no events in 2014 and in March–September 2015? The research unit can analyze the decrease in the number of events between November 2012–2013 and November 2015. Or why is there a decrease in the number of vehicles passing this site in January 2013 and 2015?

**Fig. 4.** Monthly operational data

## 4.4  An ID (Identity Card) of a Site

This screen is a drill down into a specific site with a graphical presentation of all the important information for engineering, control, and research units. The road safety parameters are demonstrated in several aspects according to desired information for the enforcement in this site. Figure 5 shows all the graphs that are shown together on the same screen, to see the "big picture". The screen is divided into four graphs: (a) speed distribution; (b) traffic count per hour; (c) the distribution of vehicles entering the junction on a red light and its speed and (d) the distribution of vehicles entering the junction on a yellow light and its speed. For each graph, a model is run in the background per site and per period to create the specific distributions from the parameters in the sensors' data log: (a) creating a speeding distribution; (b) creating a traffic count distribution; (c) creating traffic count and speeding distributions when the red traffic light is operated; (d) creating traffic count and speeding distributions when the yellow traffic light is operated;

This data can help the engineers to understand driver behavior according to the traffic light schedule and the junction's construction. From Fig. 5a we can see that there

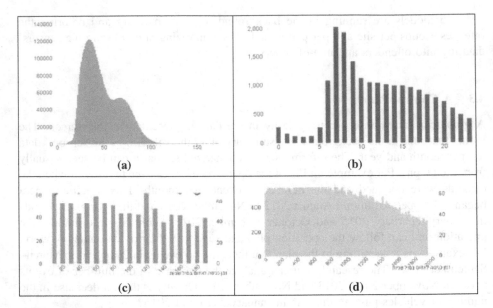

**Fig. 5.** (a) Speed distribution per site. (b) Traffic count per hour per site. (c) Number and speed of vehicles entering on a red light. (d) Number and speed of vehicles entering on a yellow light (Color figure online)

are few speed offenders at this site, and their offences are on the low fine level (up to 100 km/h) above the speed limit (90 km/h). On the other hand, there are many vehicles that cross the junction on a red light (Fig. 5c). Each figure can be easily changed to another display, for example: the traffic count (Fig. 5b) divided into days, zoom in to a specific range of the x-axis scale, or different periods.

## 4.5   A Geographic Information System Presentation

Another useful presentation is a Geographic Information System (GIS) (Fig. 6). Here, the automatic enforcement site is presented with its surroundings – other locations of traffic tickets are presented. The data can be easily filtered by time or type of ticket. The model that runs in the background is a geographic model that displays each ticket's location on the map. This presentation can help the decision maker with a decision to operate other enforcement resources in this neighborhood.

## 4.6   A Simulation Screen

This screen is a What-If screen (Fig. 7) with a friendly user interface. In this screen, the users can enter an optional enforcement level per site and the model that runs in the background calculates the new scenario. It changes the different parameters in the flow of the production line per all the operation units, i.e. the predicted percentage of drivers

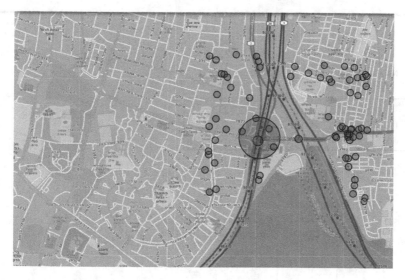

**Fig. 6.** A GIS screen

who will get a fine that will want to go to court and the predicted number of court summons that will be canceled. A graphical speed distribution (similar to Fig. 5a) is presented to help the users with the decision. In this screen the users can check different alternatives. The parameters entered by the user are combined with the databases of traffic flow, traffic distribution, and traffic tickets. A mathematical calculation is made in the background to present the information required for the final decision, i.e. the overload at the court, the difference in the number of tickets. The result is a sensitive analysis for understanding the effect of the new planned enforcement level on the number of traffic tickets per type and the load this causes the different working units in the process (encoders and traffic courts).

## 4.7   The Raw Data Screen

Figure 8a presents an example of the raw data including all the fields for a specific month and a specific site. Each raw data represents a vehicle that crosses the electromagnetic loops. The fields are: date, time, the day of the week, the time the vehicle crosses the junction in the light cycle (red-yellow-green), the vehicle's lane, the length of the vehicle and the validation of this measurement, the speed of the vehicle and the validation of this measurement, and whether the vehicle is an offender. In addition, a model runs in the background to create a pivot table, as shown in Fig. 8b, based on the raw data table. This screen is for researchers to export the raw data to statistical software for further analysis, and to help decision makers regarding optimum enforcement levels.

All the screens above have the advantages of a BI system: Default presentation for the manager level "what you see is what you get", with maximal relevant data and information on the main screens. From the operational, engineering, and research

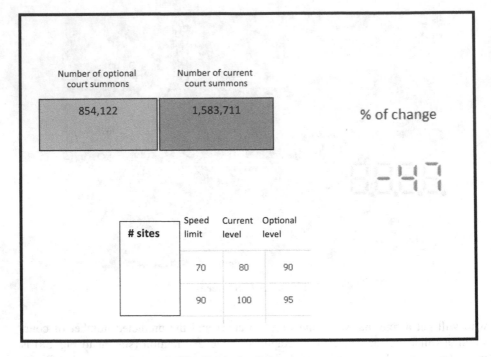

**Fig. 7.** A simulation screen

| "Date" | "Time" | "Green time" | "Yellow time" | "Red time" | "Lane nr" | "Length" | "Length validation" | "Offence type" | "Speed" | "Spd validation" |
|---|---|---|---|---|---|---|---|---|---|---|
| 2016/09/08 | 05:34:29 | 18710 | 0 | 0 | 3 | 379 "v" | | "no violation" | 31 "v" | |
| 2016/09/03 | 16:04:33 | 9310 | 0 | 0 | 3 | 379 "v" | | "no violation" | 31 "v" | |
| 2016/09/02 | 12:40:08 | 7820 | 0 | 0 | 1 | 379 "v" | | "no violation" | 31 "v" | |
| 2016/09/07 | 19:47:20 | 10130 | 0 | 0 | 1 | 379 "v" | | "no violation" | 31 "v" | |
| 2016/09/06 | 06:15:23 | 47990 | 0 | 0 | 1 | 379 "v" | | "no violation" | 31 "v" | |
| 2016/09/07 | 18:03:04 | 3470 | 0 | 0 | 2 | 379 "v" | | "no violation" | 31 "v" | |
| 2016/09/03 | 17:24:30 | 6190 | 0 | 0 | 2 | 379 "v" | | "no violation" | 31 "v" | |
| 2016/09/03 | 22:42:07 | 11440 | 0 | 0 | 2 | 379 "v" | | "no violation" | 31 "v" | |
| 2016/09/02 | 13:44:52 | 10170 | 0 | 0 | 2 | 379 "v" | | "no violation" | 31 "v" | |
| 2016/09/05 | 05:28:36 | 15130 | 0 | 0 | 2 | 379 "v" | | "no violation" | 31 "v" | |
| 2016/09/08 | 19:22:28 | 610 | 2500 | 0 | 3 | 379 "v" | | "no violation" | 49 "v" | |
| 2016/09/04 | 06:08:49 | 32360 | 0 | 0 | 1 | 379 "v" | | "no violation" | 49 "v" | |
| 2016/09/03 | 20:32:07 | 12780 | 0 | 0 | 2 | 379 "v" | | "no violation" | 49 "v" | |
| 2016/09/07 | 05:21:54 | 21380 | 0 | 0 | 1 | 379 "v" | | "no violation" | 50 "v" | |
| 2016/09/08 | 15:47:30 | 35260 | 0 | 0 | 1 | 379 "v" | | "no violation" | 50 "v" | |
| 2016/09/08 | 10:03:50 | 25430 | 0 | 0 | 2 | 379 "v" | | "no violation" | 50 "v" | |

**(a)**

| =class([...    ▾ | Passing |
|---|---|
| | **194,390** |
| 0 <= x < 10 | 25,733 |
| 10 <= x < 20 | 913 |
| 20 <= x < 30 | 14,451 |
| 30 <= x < 40 | 24,473 |
| 40 <= x < 50 | 24,155 |
| 50 <= x < 60 | 22,862 |
| 60 <= x < 70 | 23,025 |
| 70 <= x < 80 | 26,681 |
| 80 <= x < 90 | 21,755 |
| 90 <= x < 100 | 8,272 |
| 100 <= x < 110 | 1,700 |
| 110 <= x < 120 | 303 |
| 120 <= x < 130 | 45 |
| 130 <= x < 140 | 17 |
| 140 <= x < 150 | 5 |

**(b)**

**Fig. 8.** (a) The raw data screen. (b) A pivot table from the raw data

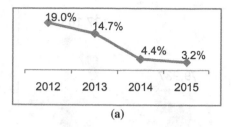
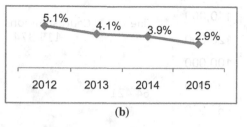

(a)                                                          (b)

**Fig. 9.** (a) The percentage of cancelled tickets. (b) The number of requests for court summons

aspects this BI system is an intuitive graphical user interface (GUI). It enables data analysis after selecting a specific parameter in a specific field, drill down to a time frame, specific location, etc. In addition, there are available bookmarks that can be saved with important drill down information for future use. With this BI system, the researcher can present the data in several displays, export it to statistical software, and merge it with other BI systems.

## 5   Results

The development of a BI system for the automatic traffic enforcement system is a huge step in the project's development from several aspects: (1) For management – to get a "big picture" of the project – total traffic passing through the project, percentage of offenders per type, severity, time, location, type of vehicle, etc.; (2) for operations, the BI system helps to determine the level of enforcement in each camera, determine the mobility of cameras between the poles and the necessity of poles; (3) helping maintenance, for example to discover if an electromagnetic loop or a camera are not working; (4) helping engineering discover sites with exceptional offenders; exceptional time frame, traffic volume, vehicle types; (5) helping researchers explore this database to check for trends, forecasting and simulation; (6) connecting this BI to other BIs of road accidents and traffic tickets to understand the "big picture".

Here are some examples that demonstrate the importance of a BI system and the improved effectiveness of the automatic traffic camera enforcement project with the BI system: (i) The process flow is now smoother for two measurements: the number of canceled tickets decreased from 19% at the beginning of the project in 2012 to 3% in 2015 (Fig. 9a), meaning fewer events entering the encoder and request units. In addition, the number of requests for court summons decreased from 5% in 2012 to 3% in 2015 (Fig. 9b), meaning a lower load on the traffic courts and request unit. (ii) More quality tickets are produced in the process. Here are some examples: the number and percentage of court summons tickets increased from 6,291 tickets as 9.8% in 2012 to 19,294 tickets as 15.2% in 2015 (Fig. 10a). The deterrence improved by shortening the time between the offence and the conviction in court (Fig. 10b).

In Fig. 10b we can see the decrease in the gap between the average number of months between the offence (the traffic ticket) and the court date. At the beginning of the project it increased dramatically to a gap of 17 months. Today, after using the BI

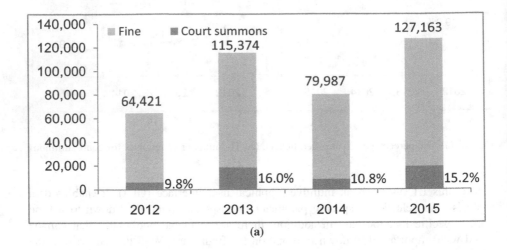

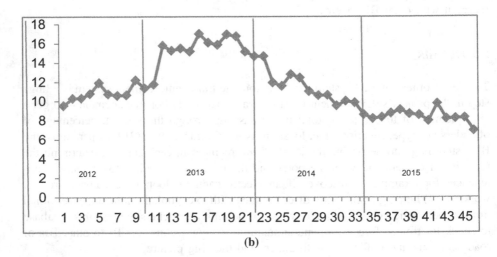

**Fig. 10.** (a)The nominal number of court summons and its percentage of the total number of tickets. (b) The average monthly gap between court date and the offence

system, it is even less than during the initial months of the project and stands at around 7 months (half of this period is a constant time of sending the ticket by mail), and with more traffic court summons than ever in the project.

The project's efficiency can be evaluated from the deterrence aspect as well. There are only 14% speed offenders (Fig. 3a) and most of them in the low fine category above the speed limit (Fig. 3b). Also, the number of red-light offenders (an offence where all offenders are caught) is reduced. In Fig. 11a we presented the number of red-light offenders for nine cameras that were operated throughout the years. The number of offenders was reduced from around 400 tickets per month to less than 300 tickets per month. In Fig. 11b, the 85th percentile of speed was calculated per 25 cameras that

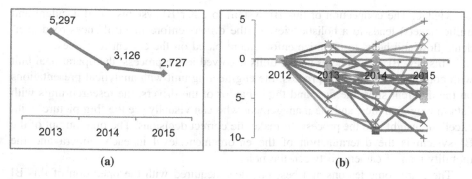

**Fig. 11.** (a) The number of red-light tickets per 9 cameras. (b) The 85th percentile of speed distribution change over the years at 25 sites (Color figure online)

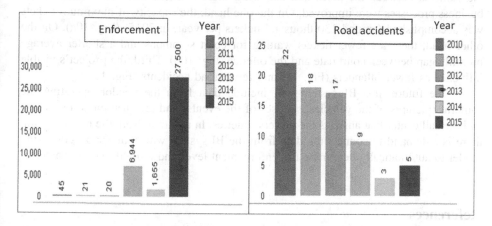

**Fig. 12.** Road accidents and traffic tickets per year per site

were operated throughout the years. In most sites, the 85th percentile of speed was reduced. In Fig. 12 a specific site is evaluated. Data on road accidents per year (on the right) and data on traffic tickets per year (on the left) are presented on the same screen. It can be seen that the number of road accidents was high until the year 2012 and then it started to decrease. One optional explanation is the installation of automatic traffic enforcement there at the end of 2012.

## 6   Summary and Further Development

In this paper we presented the stages of building a new BI system for the new automatic enforcement cameras project as PAIS. The large amount of data that is gathered each day from the sensors and the electromagnetic loops in the roads was the trigger to build this advanced BI system to efficiently and effectively operate the new automatic enforcement system. This BI system takes the data and converts it to information and knowledge. The connection of this BI system to other BI systems (road accident and traffic tickets) leads to a holistic view of the drivers-enforcement-offences-road accidents flow and helps to optimize enforcement based on the current resources.

This new BI system helps all the units involved in the process: the operational unit with control elements in the process, the engineering unit with analytical presentations on the different sites, the flow and the behavior of the drivers, the research units with data mining analytics, and the management who can visually see the 'big picture' with excellent insight into the process, to make the correct decisions. The final output of the BI system is the determination of the enforcement level in each camera and the mobility plan of cameras between the poles.

There are some lessons and best practices acquired with the operation of this BI new system. First, as in [21], the knowledge from the information affects tactical decisions for setting the enforcement speed level, and the strategic decision for moving the cameras between the poles and acquiring more cameras and poles. Second, the business processes were improved. On the one hand, the capacity of maximum tickets was accomplished (around 130 thousand tickets per year, in 2015) (Fig. 10a). On the other hand, there are fewer tickets waiting for court summons and a shorter average monthly gap between court date and the offence (Fig. 10b). Third, the project's visible influence as fewer offenders (Fig. 11) and fewer road accidents (Fig. 12).

In the future, this BI system will include data from the vendor operating the cameras (pictures of the vehicles, statuses of the events) and the customers: the courts and the call center that answers the drivers' queries. In addition, from the research unit, there is a demand to connect the data from the BI system with a linear programming model to automatically determine the enforcement level and mobility of the cameras.

# References

1. WHO (World Health Organization): Road traffic injuries. Fact sheet N°358, May 2016
2. Elvik, R.: Speed limits, enforcement, and health consequences. Ann. Rev. Health **33**, 225–238 (2012)
3. Elvik, R.: A re-parameterisation of the Power Model of the relationship between the speed of traffic and the number of accidents and accident victims. Accid. Anal. Prev. **50**, 854–860 (2013)
4. Organization for Economic Co-operation and Development (OECD): Speed management (No. 55921). Organization for Economic Co-operation and Development, Paris (2006)
5. Elvik, R.: Speed enforcement in Norway: testing a game-theoretic model of the interaction between drivers and the police. Accid. Anal. Prev. **84**, 128–133 (2015)
6. De Pauw, E., Daniels, S., Brijs, T., Hermans, E., Wets, G.: An evaluation of the traffic safety effect of fixed speed cameras. Saf. Sci. **62**, 168–174 (2014)
7. Aarts, L., van Schagen, I.: Driving speed and the risk of road crashes: a review. Accid. Anal. Prev. **38**, 215–224 (2006)
8. Wijers, P.J.: Automated enforcement, get it right, make it safe. In: 16th Road Safety on Four Continents Conference Beijing, China, 15–17 May 2013
9. Gatso. http://www.gatso.com/en/about-gatso/history/. Accessed 23 May 2016

10. Işık, O., Jones, M.C., Sidorova, A.: Business intelligence success: the roles of BI capabilities and decision environments. Inf. Manag. **50**(1), 13–23 (2013)
11. Bahrami, M., Arabzad, S.M., Ghorbani, M.: Innovation in market management by utilizing business intelligence: introducing proposed framework. Procedia - Soc. Behav. Sci. **41**, 160–167 (2012)
12. Negash, S.: Business intelligence. Commun. Assoc. Inf. Syst. **13**, 177–195 (2004)
13. Chen, H., Chiang, R.H.L., Storey, V.C.: Business intelligence and analytics: from big data to big impact. MIS Q. **36**(4), 1165–1188 (2012)
14. Elbashir, M.Z., Collier, P.A., Davern, M.J.: Measuring the effects of business intelligence system: the relationship between business process and organizational performance. Int. J. Acc. Inf. Syst. **9**, 135–153 (2008)
15. Cody, W.F., Kreulem, J.T., Krishna, V., Spangler, W.S.: The integration of business intelligence and knowledge management. IBM Syst. J. **41**(4), 697–713 (2002)
16. Nenortaite, J., Butleris, R.: Improving business rules management through the application of adaptive business intelligence technique. Inf. Technol. Control **38**(1), 21–28 (2009)
17. Aalst, Wil M.P.: Process-Aware Information systems: lessons to be learned from process mining. In: Jensen, K., Aalst, Wil M.P. (eds.) Transactions on Petri Nets and Other Models of Concurrency II. LNCS, vol. 5460, pp. 1–26. Springer, Heidelberg (2009). doi:10.1007/978-3-642-00899-3_1
18. Weber, B., Reichert, M., Rinderle-Ma, S.: Change patterns and change support features – enhancing flexibility in process-aware information systems. Data Knowl. Eng. **66**, 438–466 (2008)
19. Reichert, M., Weber, B.: Enabling Flexibility in Process-Aware Information Systems, Challenges, Methods. Technologies. Springer, Heidelberg (2012)
20. Dumas, M., Van der Aalst, W.M., Ter Hofstede, A.H.: Process-Aware Information Systems: Bridging People and Software Through Process Technology (2005). ISBN: 978-0-471-66306-5
21. Watson, H.J., Wixom, B.H., Hoffer, J.A., Anderson-Lehman R., Reynolds, A.M.: Real-time Business Intelligence: Best Practices at Continental Airlines Information Systems Management (2006). Accessed 21 Dec 2006
22. Trieu, V.H.: Getting value from business intelligence systems: a review and research agenda. Decis. Support Syst. **93**, 111–124 (2017)
23. CBS (Central Bureau of Statistics): Press release 085/2016 (2016). http://www.cbs.gov.il/reader/newhodaot/hodaa_template.html?hodaa=201627085

# Multicriteria Decision Making for Healthcare Facilities Location with Visualization Based on FITradeoff Method

Marta Dell'Ovo[1(✉)], Eduarda Asfora Frej[2], Alessandra Oppio[3], Stefano Capolongo[1],
Danielle Costa Morais[2], and Adiel Teixeira de Almeida[2]

[1] Department of Architecture, Built Environment and Construction Engineering,
Politecnico di Milano, via Bonardi 9, 20133 Milan, Italy
marta.dellovo@polimi.it
[2] Center for Decision Systems and Information Development – CDSID,
Federal University of Pernambuco – UFPE, Cx. Postal 7462, Recife, PE 50630-970, Brazil
[3] Department of Architecture and Urban Studies, Politecnico di Milano,
via Bonardi 3, 20133 Milan, Italy

**Abstract.** This paper proposes an application of the Flexible Interactive Tradeoff (FITradeoff) method for siting healthcare facilities. The selection of the location of complex facilities, as hospitals, can be considered as a multidimensional decision problem for the several issues to be taken into account and, moreover, for the variety of stakeholders that should be involved. The case study under investigation is the location of "La Città della Salute", a new large healthcare facility in Lombardy Region (Italy). Starting from a cross disciplinary literature review, a multidimensional evaluation framework has been defined and applied to the case study by considering the point of view of one Decision Maker (DM). The application shows that a smaller effort is required from the DM using the FITradeoff method.

**Keywords:** Healthcare facility location · MCDM · FITradeoff · Decision-making · Visualization

## 1 Introduction

The decision-making is a complex process and a complex skill to learn for the several issues to consider and because "there are many psychological traps that can cause our thinking to go astray" [1]. There are several steps to consider in order to structure decisions. In particular, Sharifi and Rodriguez (2002) [2] organize the Planning and Decision-Making Process in three different phases: Intelligence (Process model); Design (Planning model); Decision/Choice (Evaluation model). The first phase considers the identification of the state of the decision and the definition of main objectives, the second phase deals with the formulation of the model and the generation of alternatives, while the last phase concerns the evaluation of alternatives and the communication of the result to the Decision Maker (DM) using an appropriate method.

The location of healthcare facilities involves many aspects, both qualitative and quantitative, and the definition of values elicitation is not easy due to several

© Springer International Publishing AG 2017
I. Linden et al. (Eds.): ICDSST 2017, LNBIP 282, pp. 32–44, 2017.
DOI: 10.1007/978-3-319-57487-5_3

stakeholders that are generally involved in these kind of choices. Each stakeholder could have a specific objective to achieve not shared by all. In order to make better decisions, it is important to select criteria able to affect the decision problem and all the categories of stakeholders that influence it, clarifying if their involvement is direct or indirect.

The World Health Organization defines health as 'a state of complete physical, mental and social well-being and not merely the absence of disease or infirmity' (WHO, 1987) [3]. The location of facilities plays a crucial role in improving the efficiency of health system [4]. Thus well-being, health and location of hospitals are directly connected by a reciprocal relationship.

Since "locating hospitals is a process that must take into consideration many different stakeholders" [5] and their satisfaction plays a key role, this paper will investigate the preference of one specific decision-maker using the Flexible Interactive Tradeoff (FITradeoff) method outlining location strategies in order to solve the decision problem.

## 2  Location of Healthcare Facilities

Many scholars have investigated the topic related to the location of healthcare facilities by applying different methodologies to specific case studies in order to solve the decision-problem, especially in the last 15 years. Noon and Hankins [6] aimed to support the decision for the location of a Neonatal Intensive Care Unit in West Tennessee displaying data for each facility, for patient density and market share supported by Geographic Information System (GIS). For Danskin and Dean [4] the location modeling of healthcare facilities takes a great importance because a wrong siting could increase death and disease, and the accessibility, defined as the ability to reach healthcare facilities, is fundamental. Murad [7], in order to plan public general hospitals at Jeddah City, used spatial data, as the road network, the hospital locations and the population coverage, and non-spatial data, as hospitals size (capacity), number of people living in each district of the city, and the population density of that district. Vahidnia et al. [8], in order to select a site for a new hospital in the Tehran urban area (Iran), has combined Geographic Information System (GIS) and Fuzzy Analytical Hierarchy Process (FAHP) into the following thematic maps: (a) the locations of five possible sites, (b) distance from arterial streets, (c) travel time area to access existing hospitals, (d) contamination, (e) land cost, (f) population density. Soltani and Marandi [9], by using Fuzzy Analytical Network Process Systems with Geographic Information System (GIS), reported the location of a hospital within the Region 5 of Shiraz metropolitan area (Iran). During the phase aimed to define alternatives, the criteria considered are: (a) distance to arterials and major roads; (b) distance to other medical centers; (c) population density and (d) parcel size. Burkley et al. [5] provided, as case study, the hospital location in four southern US states, to measure different degrees of the accessibility. He argued that there are several factors that can influence it: (a) existing facilities are not well located, (b) there are fewer facilities, (c) the road or transportation network might be inefficient, (d) the distribution of the population is difficult to serve. Wu et al. [10], considering the complex nature of factors involved in understanding the degree of concentration of medical resources in the Taipei area, combined the Grey Relational Clustering (GRC) method and a hierarchical clustering analysis. The analysis defined for each hospital concerns (a) year, (b) bed number, (c) hospital property (private or public) and (d) distribution district. According to Faruque et al. [11], there are several geographical factors that can negatively influence health outcomes. Their analysis is focused on underlining the distribution of patients with diabetes and their distance to existing healthcare facilities with the aim to define unserved areas. Chiu and Tsai [12] aimed to find the best way to expand medical services using the Analytical Hierarchy Process (AHP) and dividing the problem in five criteria, in turn composed by sub-criteria. The criteria analyzed are: (a) demand of medical services, based on population demand and density, (b) construction cost, considering construction cost and availability of land, (c) transportation, including private and public transport and parking, (d) sector support, to evaluate administrative departments and sectors involved in the process, (e) future developments that considers existing hospitals and possible expansion plans. Abdullahi et al. [13] compared two weighting techniques, the Analytical Hierarchy Process (AHP) and the Ordinary Least Square (OLS) for the construction of hospitals in the Qazvin city (Iran). In particular three districts have been explored and variables analyzed in the comparison process are: (a) Distance to squares, (b) Distance to fire stations, (c) Distance to taxi stations, (d) Distance to bus stations, (e) Distance to noisy areas, (f) Distance to polluted areas and (g) Distance to populated points. Moreover, technical, environmental and socioeconomic issues have been considered and constraints that could not allow developments. Lee and Moon [14] in the study city of Seoul (South Korea) collected spatial information from Statistics Korea and its Statistical Geographical Information System (SGIS) and applied a socio-demographic analysis to investigate characteristics of different areas. The variables defined are: (a) number of hospitals, (b) total population, (c) population aged > 65 years, (d) number of businesses, (e) number of workers, (f) road area, (g) number of subway entrances and (h) residential area. Another important issue useful to consider in the location of services is the traffic. Beheshtifar and Alimoahmmadi [15] combined geographical information system (GIS) analysis with a multiobjective genetic algorithm to solve the location-allocation of healthcare facilities. The study area selected to apply the methodology is located in region 17 of Tehran (Iran) and four objectives have been pointed out: travel costs and the cost of land acquisition and facility establishment, and to maximize accessibility and suitability of the selected sites. Zhang et al. [16] used GIS and spatial accessibility indexes for the spatial analysis of rural medical facilities. Data collected deals with the number of technical staff, the number of hospital beds and the number of large medical equipment of hospitals in the area under investigation (China). According to Dun and Sun [17] the location planning is a crucial issue for improving patient satisfaction, and it is defined by 4 indexes: (a) response time; (b) patient wait time; (c) service quality; (d) service price. Kim et al. [18] conducted an analysis based on evidence-based decision-support system, supported by GIS for the hospital site selection of aging people. Considering their research, the success of healthcare facilities depends on its location. Criteria analyzed are divided in three categories. (a) Needs (demographics, socio-economics, health conditions, existing health services, health services utilization). (b) Capacity (infrastructure: construction phase, infrastructure: accessibility, infrastructure: other utilities) (c) Support (financial support, community support).

From the analysis of the literature review emerges a convergence of criteria, even if analyzed with different disciplinary prospective and methodologies. In fact, this investigation has been useful in order to classify and select crucial criteria to select the most suitable site to locate healthcare facilities.

## 3   The FITradeoff Method

In the context of Multiattribute Value Theory (MAVT), multiple criteria decision problems are solved by scoring the alternatives straightforwardly, through a weighted sum. Regarding this approach, one of the main issues to concern about is the elicitation of the scaling constants of the criteria, since these parameters do not represent only the relative importance of the criteria, but a scaling factor is also involved.

This is probably what has enhanced the development of many procedures for eliciting scaling constants in the additive model, some of them discussed by Riabacke et al. (2012) [19]. The tradeoff procedure, proposed by Keeney & Raiffa (1976) [20], has a strong axiomatic structure [21], but it is very difficult for the decision maker to answer the questions required [22], since he/she has to specify an exact value that makes two consequences indifferent, which is not an easy task. The decision maker may not be able to answer these questions in a consistent way, and according to behavioural studies [23], this procedure presents about 67% of inconsistencies when applied.

The flexible interactive tradeoff (FITradeoff), proposed by de Almeida et al. (2016) [24], is a method for elicitation of criteria weights in additive models. This new method is based on the tradeoff procedure, keeping its axiomatic structure, but improving its applicability for the DM, with cognitively easier questions in the elicitation process.

The FITradeoff method uses partial information about the decision maker's preferences to elicit scaling constants and so to determine the most preferred alternative of a specified set, according to an additive model [24]. Since it works with partial information, less information is required from the decision maker, and the information required is easier to provide. Thus, since the FITradeoff generates results similar to those of the classical tradeoff procedure, the expectation is that the rate of inconsistencies will be lower because the DM needs to make less cognitive effort as only partial information is required [24], but future behavioural studies works should focus on these inconsistences.

The elicitation process is conducted in a flexible and interactive way: the DM answers to questions about strict preference relations - which are easier compared to the indifference point required in the tradeoff procedure - and the FITradeoff Decision Support System (DSS) systematically evaluates the potentially optimal alternatives, through linear programming problems (LPP). The process finishes either when a unique solution is found or when the decision maker is not willing to give additional information. The FITradeoff DSS allows the DM to see partial results that can be displayed graphically at any interaction during the process, in a flexible way. With these results, the DM can choose whether or not to continue the process, depending on whether the partial results are already sufficient for his/her purpose. The FITradeoff DSS works as illustrated in Fig. 1.

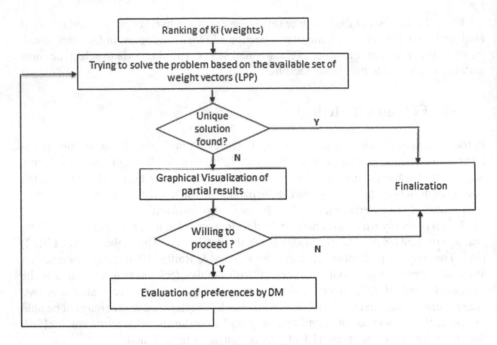

**Fig. 1.** The FITradeoff process

The first step is the ranking of criteria. The ranking now becomes a constraint for the LPP model. If a unique solution is found considering the current weight space, the process finishes at this point. Otherwise, the DM will be able to see the partial results through visualization analysis and compare the remaining alternatives, in order to decide if he/she wants to continue the process or not. If the DM is not willing to proceed, the elicitation stops and he can use the partial results provided by the DSS to make a decision. Otherwise, the DM's preferences are evaluated through questions asking about preference relations between consequences. The answer is added to the LPP model as another constraint, tightening the weight space. And so the process goes on, until a unique solution is found or DM is not willing to proceed.

This flexible process asks the decision maker only the information strictly required, and so it avoids overloading the DM with a large amount of questions. Besides that, the questions asked are easier, saving time and effort from the decision maker.

## 4   Case Study

### 4.1   La Città Della Salute

The project, "La Città della Salute" in the city of Milan (North of Italy), has been proposed in the early 2000s with the aim of relocating two existing healthcare facilities, the Istituto Neurologico Carlo Besta and the Istituto dei Tumori. The aim of the program was to answer to scientific and cultural changes of contemporary medicine by combining in a unique pole healthcare services focused and specialized on research, teaching, science and training. The process to select the area has not been supported by a specific Decision Support System (DSS) for evaluating the suitability of the site and six different locations have been proposed during years (Perrucchetti, Ortomercato, Rogoredo, Sacco, Expò and Sesto San Giovanni). In 2013 the future development of the area of Sesto San Giovanni has been defined by Renzo Piano, according to the project guidelines, called Metaprogetto, defined by the Health Ministry with a team of experts in 2000. In this case, political and economic reasons prevailed on the technical ones, and after 3 years the project did not start yet and people living close to the two existing hospitals are concerned about buildings abandoned after their relocation.

## 4.2  Criteria Definition

The criteria selected to locate "La Città della Salute" and considered as crucial for the analysis of a site, are the result of the literature review. Given the multidimensional nature of the decision problem, a multi-criteria evaluation framework has been defined. It is composed by four macro areas and each of them by four criteria, in particular [25]: 1. Functional Quality divided in 1.1 Building density that considers population living in the area where the site is located while the 1.2 Health demand the population over 65 years old, 1.3 Reuse of built-up area that takes into account if the site is already developed and 1.4 Potential of the area to become an attractive pole to create a new center with a positive effect on the neighboring community. The 2. Location Quality includes 2.1 Accessibility to consider private, public and alternative transports and the availability of parking lots and number of accesses, 2.2 Existing hospital to evaluate the presence of hospitals nearby the site under investigation that could be both a positive and a negative aspect according to healthcare facility to locate, 2.3 Services to calculate the number of facilities in the surrounding areas according to specific categories (libraries, schools, churches, post offices, fire stations, restaurants, etc.), 2.4 Sewer system to choose a site close to an efficient sewerage to avoid hygiene problems. The 3. Environmental Quality is composed by the 3.1 Connection to green areas [26] calculated according to the compliance to specific and defined parameters, 3.2 Presence of rivers and canals to avoid hydraulic and hydrological instability, 3.3 Air and noise pollution investigated according to the presence of noise sources and pollutants such us PM10, NO2 and O3, 3.4 Land contamination to assess the suitability of the area considering if further interventions are necessary, as the reclamation, to locate there the hospital. The last macro-area considers 4. Economic Aspects, in fact, 4.1 Land size and shape allows to define the flexibility of the area, 4.2 Land ownership, if it is private or public, 4.3 Land cost, its monetary value per sqm and 4.4 Land use to develop the area in an ethical way.

## 4.3  Stakeholders Analysis

The decision problem can be modelled by the involvement of several categories of stakeholders into the decision-making process. In fact, according to the stakeholder classification provided by Dente (2014) [27], it is possible to consider five different actors able to influence the decision problem with their preference and interests. Since choices about healthcare location are characterized by complex political and technical aspects and high level of uncertainty due to the duration of bureaucratic and administrative procedures, several stakeholders with different interest should be taken into consideration. It is also important to define how the choice of one actor can affect other actors and which is the role they play in a decision-making process. According to the aim of this paper it is useful to define for each category of stakeholders their interest and to recognize and detect their position in the decision problem in order to define their preference (Table 1).

**Table 1.** Identification of stakeholders.

| Category | Interest | Actors |
|---|---|---|
| Political actors | Represent citizens | Health and Urban Councillor |
| Bureaucratic actors | Formal competence | Health and Urban General Manager |
| Actors with special interests | Sustain cost or benefits | Local Health Unit Director |
| Actors with general interests | Represent who is not able to defend themselves | NGO, NPO, Common people |
| Experts | Specific knowledge | Architects, Urban planners, Technology expert, Doctor, Transport policy expert, Urban economist, Project appraisal expert, Environmental hygiene expert, Management engineer, Environmental practice expert |

Once analyzed them, the category 'expert' has been selected to act as the DM, to test and apply the FITradeoff elicitation method, because considered as the most appropriate to understand and solve the decision problem for its competences and skills on this field. In particular, an architect/urban planner with knowledge about healthcare facilities requirements and urban problems has been questioned also in the first step of the process to assess weighs to criteria and rank them.

## 5    Application and Results

### 5.1    Application

The FITradeoff elicitation method has been applied to solve the healthcare facility location problem described above. The first step was the ranking of criteria weights, which was made by pairwise comparison in the DSS. The resulted order was, in descending order:

C1 - Accessibility
C2 - Services
C3 - Presence of rivers and canals

C4 - Air and Noise Pollution
C5 - Land Cost
C6 - Sewerage system
C7 - Land contamination
C8 - Existing hospital
C9 - Reuse of built-up areas
C10 - Building density
C11 - Health demand
C12 - Land ownership
C13 - Land use
C14 - Connection to green areas
C15 - Potential of area to become attractive point
C16 - Land size and shape.

After the ranking of criteria weights, the LPP model resulted on three non-dominated alternatives, which were Perruchetti, Ortomercato and Sesto San Giovanni. A unique solution was not found yet, so the DM could proceed, at this stage, the visualization analysis of the partial results. The FITradeoff DSS provides graphical visualization as shown in Fig. 2.

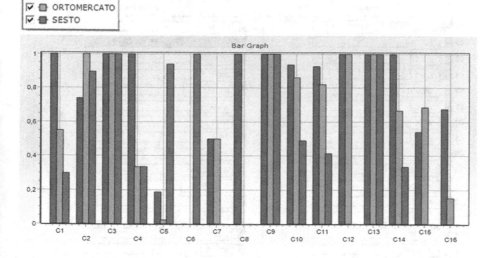

**Fig. 2.** Bar graphic of the partial results, after the ranking of criteria weights.

This graphic represents the values of the alternatives in each criterion, normalized in 0–1 scale. The criteria are ordered from left to right. It can be observed, for instance, that against the first criterion (C1 - Accessibility), the performance of Ortomercato is approximately half of the one of Perrucchetti, while Sesto has the lowest value. With respect to the second criterion, however, Ortomercato has a slight advantage over the other alternatives. About C3 - Presence of rivers and canals, the alternatives are all tied.

In the rank based on C4 - Air and Noise Pollution, Perrucchetti has a great advantage, with a value almost three times greater than the value of Ortomercato and Sesto, which are tied. This analysis could be developed for all the criteria, so that the decision maker can compare the values of the alternatives that are potentially optimal for the problem in an illustrative way. With this analysis, if the DM thinks that the information provided is already enough to aid the decision, the elicitation process can be stopped. Otherwise, the process can be reiterated by going on to answer to the questions.

Looking at these partial results, the DM has decided to continue the elicitation process and start to answer the questions. Figure 3 illustrates an example of question made for the decision maker by the DSS, by using a visualization tool. The DM has to choose whether he prefers consequence A or consequence B. Consequence A has the first criterion (Accessibility) with an intermediate performance (blue bar) calculated by the DSS according to a heuristic [24], and all the others with the worst performance (red bars). Consequence B has the second criterion (Services) with the best performance (green bar), and all the others with the worst performance. The answer will lead us to an inequality which relates the values of the scale constants of Accessibility and Services. This inequality is then incorporated to the LPP model as a restriction.

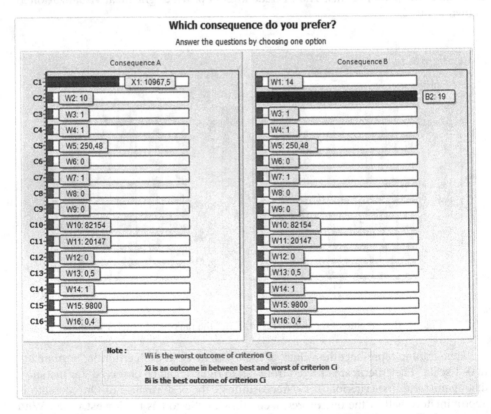

**Fig. 3.** Question made for the DM during the elicitation. (Color figure online)

After the first question answered, the set of non-dominated alternatives remained the same. After four questions answered, Ortomercato has been dominated.

An analysis similar to that of Fig. 2 can be made at this stage, by comparing only Perrucchetti and Sesto San Giovanni. As the DM was still willing to continue the process, a unique solution has been found after six question answered. Sesto San Giovanni has been dominated and Perrucchetti chosen.

FITradeoff works with LPP models, where the constraints are inequalities gathered from the answers given by the DM. With these inequalities, a weight space is obtained. It means that the final solution remains the optimal one for the whole set of weights in the area defined by the weight space resulted from the answers of the decision maker. The weight space can be visualized in Fig. 4, with the maximal and minimum value for each scaling constant represented by the dots, connected by lines.

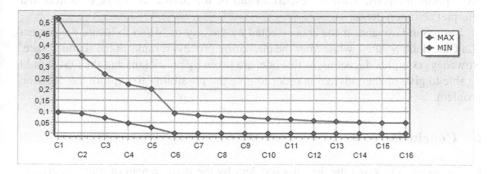

**Fig. 4.** Final weight space

By applying FITradeoff to solve this problem, a solution has been found with only 6 questions answered by the DM. If the traditional tradeoff procedure was applied, it would require at least 15 questions (Number of criteria - 1), to solve the equation system and find the values of the scaling constants. Besides that, the kind of information required in FITradeoff is easier to provide and requires less cognitive effort from the DM. Another advantage of this approach is the flexibility of the method, which makes it possible for the DM to stop the elicitation process, if the results displayed by the graphics are already enough for the evaluation's purposes. This would reduce even more the number of questions, saving time and effort from the decision maker.

## 5.2   Discussion of the Results

The interaction with the DM has revealed the flexibility of the method, easy to be used and understood. In fact, the DM has been guided in the process in order to get the meaning of different criteria without misunderstanding or personal interpretation, and once understood them it has been easier to assess their importance. As it emerges from the graph obtained from the software (Fig. 2), the area resulted as the most suitable to locate healthcare facilities (Perrucchetti), ranks also as the best one in most of the criteria selected or at least in the most important. It means that the DM has a strong influence in defining the final result, ordering criteria according to his/her preferences and interests. In fact, changing the order and assigning more importance to other aspects, the result could change.

Different from other multicriteria methods, FITradeoff does not find exact values for the scale constants, but a whole weight space for which the chosen alternative (in this case, Perrucchetti) dominates all the others. So talking about sensitivity analysis regarding the values of scale constants doesn't make much sense, since the space of weights in Fig. 4 already shows how sensitive is the winner alternative with respect to the values of the weights.

The "expert" has been chosen as DM, because considered as the most objective one and with specific knowledge [27] regarding the location of healthcare facilities. In particular other stakeholders involved in the decision making process could have their own special interests, while the expert should be the person in charge to collect and interpret common needs.

Sesto San Giovanni is the area selected by the city of Milan to locate the project "La Città della Salute", while after the elicitation process it ranks at the second place. Nowadays complex decisions, as this one, are taken by politicians and the DM should be able to give his/her advices in order to give the possibility to explore the decision problem.

## 6    Conclusions

Exploring and modeling the decision problem by the involvement of multiple criteria allows to consider all the aspects necessary in order to avoid future problems related to the choice of the site. Moreover, the tool improves the transparency of the decision-making process and simplifies and speeds up its procedures [28].

From the analysis of the literature review it is clear that many scholars are trying to structure and solve this complex decision problem. It means that it is a topic of international interest and the definition of an evaluation framework could provide benefits under different point of view. Social: the location takes into account the population density and the health demand, promoting also the reuse of built-up areas and then their requalification. Environmental: the sustainability is one of the most important aim to pursue. Economic: the evaluation tool reduces the duration of selection procedures. The FITradeoff elicitation method has been applied to solve the healthcare facility location problem analyzed here. The FITradeoff DSS supports the decision process in a flexible way, by providing visualization tools for elicitation questions and analysis of results. The partial results can be analyzed and the process stopped at any time, according to the achievement of DM's preferences. The elicitation process for the current case study has been completed with only 6 questions answered, which shows that a smaller effort is required from the DM, compared to other elicitation methods.

The application of the methodology on real case studies is the first step to demonstrate the efficiency of the tool that could be constantly implemented and updated in order to be used in international and cross-disciplinary contexts. A possible future development of the methodology could be the interaction of different stakeholders in the decision making process. The statement of the power of their contribution to solve the decision problem it is not an assumption but it has to be evaluated in order to take into consideration their positions and interests, often conflictual. In fact, in real world case studies, there are several point of views to deal with and to collect, with the aim of satisfying different expectations. FITradeoff method could be easily adapted for Group Decision problems, by conducting the elicitation with several stakeholders and then comparing and discussing information to obtain a shared result.

The inclusion of different stakeholders, with respect to the decision problem, could improve the decision modeling step and the robustness of the results.

**Acknowledgments.** This study was partially sponsored by the Brazilian Research Council (CNPq) for which the authors are most grateful.

# References

1. Keeney, R.L.: Make better decision makers. Decis. Anal. **1**(4), 193–204 (2004)
2. Sharifi, M.A., Rodriguez, E.: Design and development of a planning support system for policy formulation in water resource rehabilitation. J. Hydroinformatics **4**(3), 157–175 (2002)
3. Capolongo, S., Buffoli, M., Oppio, A.: How to assess the effects of urban plans on environment and health. Territorio **73**, 145–151 (2015)
4. Daskin, M.S., Dean, L.K.: Location of Health Care Facilities, pp. 43–76. Kluwer Academic Publisher, Boston (2004)
5. Burkey, M.L., Bhadury, J., Eiselt, H.A.: A location-based comparison of health care services in four U.S. states with efficiency and equity. Socio-Econ. Plan. Sci. **46**, 157–163 (2012)
6. Noon, C.E., Hankins, C.T.: Spatial data visualization in healthcare: supporting a facility location decision via gis-based market analysis. In: 34th Hawaii International Conference on System Sciences (2001)
7. Murad, A.A.: Using GIS for planning public general hospitals at Jeddah City. Environ. Des. Sci. **3**, 3–22 (2005)
8. Vahidnia, M.H., Alesheikh, A.A., Alimohammadi, A.: Hospital site selection using fuzzy AHP and its derivatives. J. Environ. Manag. **90**, 3048–3056 (2009)
9. Soltani, A., Marandi, E.Z.: Hospital site selection using two-stage fuzzy multi-criteria decision making process. J. Urban Environ. Eng. **5**, 2–43 (2011)
10. Wu, C.R., Lin, C.T., Chen, H.C.: Optimal selection of location for Taiwanese hospitals to ensure a competitive advantage by using the analytic hierarchy process and sensitivity analysis. Build. Environ. **42**, 1431–1444 (2007)
11. Faruque, L.I., Ayyalasomayajula, B., Pelletier, R., Klarenbach, S., Hemmelgarn, B.R., Tonelli, M.: Spatial analysis to locate new clinics for diabetic kidney patients in the underserved communities in Alberta. Nephrol. Dial. Transplant. **27**, 4102–4109 (2012)
12. Chiu, J.E., Tsai, H.H.: Applying analytic hierarchy process to select optimal expansion of hospital location. In: 10th International Conference on Service Systems and Service Management, pp. 603–606 (2013)
13. Abdullahi, S., Mahmud, A.R., Pradhan, B.: Spatial modelling of site suitability assessment for hospitals using geographical information system-based multicriteria approach at Qazvin city. Iran Geocarto Int. **29**, 164–184 (2013)

14. Lee, K.S., Moon, K.J.: Hospital distribution in a metropolitan city: assessment by a geographical information system grid modelling approach. Geospatial Health **8**, 537–544 (2014)
15. Beheshtifar, S., Alimoahmmadi, A.: A multiobjective optimization approach for location-allocation of clinics. Int. Trans. Oper. Res. **22**, 313–328 (2015)
16. Zhang, P., Ren, X., Zhang, Q., He, J., Chen, Y.: Spatial analysis of rural medical facilities using huff model: a case study of Lankao county, Henan province. Int. J. Smart Home **9**, 161–168 (2015)
17. Du, G., Sun, C.: Location planning problem of service centers for sustainable home healthcare: evidence from the empirical analysis of Shanghai. Sustainability **7**, 15812–15832 (2015)
18. Kim, J.I., Senaratna, D.M., Ruza, J., Kam, C., Ng, S.: Feasibility study on an evidence-based decision-support system for hospital site selection for an aging population. Sustainability **7**, 2730–2744 (2015)
19. Riabacke, M., Danielson, M., Ekenberg, L.: State-of-the-art prescriptive criteria weight elicitation. Adv. Decis. Sci. **2012**, 1–24 (2012)
20. Keeney, R.L., Raiffa, H.: Decision Analysis with Multiple Conflicting Objectives. Wiley & Sons, New York (1976)
21. Weber, M., Borcheding, K.: Behavioral influences on weight judgments in multiattribute decision making. Eur. J. Oper. Res. **67**(1), 1–12 (1993)
22. Edwards, W., Barron, F.H.: SMARTS and SMARTER: improved simple methods for multiattribute utility measurement. Organ. Behav. Hum. Decis. Process. **60**, 306–325 (1994)
23. Borcherding, K., Eppel, T., Von Winterfeldt, D.: Comparison of weighting judgments in multiattribute utility measurement. Manag. Sci. **37**(12), 1603–1619 (1991)
24. de Almeida, A.T., Almeida, J.A., Costa, A.P.C.S., Almeida-Filho, A.T.: A new method for elicitation of criteria weights in additive models: flexible and interactive tradeoff. Eur. J. Oper. Res. **250**(1), 179–191 (2016)
25. Oppio, A., Buffoli, M., Dell'Ovo, M., Capolongo, S.: Addressing decisions about new hospitals' siting: a multidimensional evaluation approach. Ann. Ist. Super Sanità **52**(1), 78–87 (2016)
26. D'Alessandro, D., Buffoli, M., Capasso, L., Fara, G.M., Rebecchi, A., Capolongo, S.: Green areas and public health improving wellbeing and physical activity in the urban context. Epidemiol Prev. **39**(5 Suppl 1), 8–13 (2015)
27. Dente, B.: Understanding Policy Decisions. SpringerBriefs in Applied Sciences and Technology, PoliMI SpringerBriefs. Springer, New York (2014)
28. Dell'Ovo, M., Capolongo, S.: Architectures for health: between historical contexts and suburban areas. Tool to support location strategies. Techne **12**, 269–276 (2016)

# Business Process Modelling and Visualisation to Support e-Government Decision Making: Business/IS Alignment

Sulaiman Alfadhel[1]([✉]), Shaofeng Liu[1], and Festus O. Oderanti[2]

[1] Plymouth Graduate School of Management,
Plymouth University, Plymouth, UK
{sulaiman.alfadhel,shaofeng.liu}@plymouth.ac.uk
[2] Hertfordshire Business School, University of Hertfordshire, Hatfield, UK
F.Oderanti@Herts.ac.uk

**Abstract.** Alignment between business and information systems plays a vital role in the formation of dependent relationships between different departments in a government organization and the process of alignment can be improved by developing an information system (IS) according to the stakeholders' expectations. However, establishing strong alignment in the context of the eGovernment environment can be difficult. It is widely accepted that business processes in the government environment plays a pivotal role in capturing the details of IS requirements. This paper presents a method of business process modelling through UML which can help to visualise and capture the IS requirements for the system development. A series of UML models have been developed and discussed. A case study on patient visits to a healthcare clinic in the context of eGovernment has been used to validate the models.

**Keywords:** IS requirements · Process modelling and visualisation · Goal modelling · UML · Requirements elicitation

## 1 Introduction

The trend toward the globalization of the business organization environment remains unabated and has generated profound renovations, both internal and external, as the 'majority of organizations seek to re-align or re-create their value chain' while endeavoring to forge closer associations with their consumers and business partners. The term eGovernment refers to the use of information system (IS) services by government agencies that have the potential to transform relationships with industry, citizens and other arms of government [20]. IS technologies can serve a variety of different ends, such as better government services to citizens, enhanced interactions between government and business industry and the management of government administration [20, 22].

In response to or in anticipation of changes in their environment, the majority of organizations are deploying information systems for this purpose at a growing rate [21, 26]. Consequently, this has raised a primary question fundamental to the current

© Springer International Publishing AG 2017
I. Linden et al. (Eds.): ICDSST 2017, LNBIP 282, pp. 45–57, 2017.
DOI: 10.1007/978-3-319-57487-5_4

business paradigm: how can a business organization actually justify its information system investments in the context of contributing to business performance, be it in terms of efficiency, amplified market share, productivity or other indicators of organizational usefulness [13–15]?

However, the process of managing and providing IS services for any government is a difficult job, due to rapid changes in the government environment and a lack of alignment between various government departments and the IS department. Strong alignment between IS and other departments of government can achieve better administration and organizational performance in many ways, such as strategic, social, cultural and structural performance [1, 5, 12].

The literature shows that researchers have studied alignment in different contexts, for example, the strategic differences between business and information systems, the structural differences between business and information systems, and the cultural differences between business and information systems [2–4]. One way of establishing strong alignment between IS and other government agencies is to develop IS in accordance with government expectations to ensure a system that meets the government's needs [10–12].

System requirements engineering is the process of determining the government's perspective on the system, which helps developers design a system that accurately meets the government's needs. To obtain accurate system requirements, it is important to understand the business environment in which the system will be deployed. Therefore, business process modelling in the context of obtaining system requirements is required prior to developing the system [7, 16, 27]. This paper presents a method of modelling and analyzing business processes with the aim of deriving the system requirements from the business goals and developing a suitable e-health system in the context of eGovernment.

## 2   Related Work

Alignment is a procedure where two different groups, business and information system, are interconnected with each other and where information system aims to provide services at all levels of the business organization in order that the business goals and objectives of the organization are attained [7, 27]. However, alignment is not a single entity; it is a vague and ambiguous process that contains of several stages, each stage representing a precise part of the business organization. Moreover, literature shows that modeling business goals in context of alignment between business and information systems and e-government has not been addressed [25]. Only a few researchers have look at this issue in context of business goal modelling, together producing a good deal of work which include: I* [31], GOMS [33], Goal-based Workflow [32], and KASO [30].

The I* approach is based on a goal modelling in the context of eGovernment idea that enables information system analysts to inspect the business requirements at an early stage of system development. It needs the organization's actors to explain the business goals and business strategy. In this modelling approach, business organizational actors have their predefine business goals and beliefs and every actor is associated to one another [31]. The GOMS and Goal-based Workflow techniques propose a goal modelling methods for system requirements elicitations in regard to illuminating the business organizational objectives and to thoughtful the current organizational situation [32, 33]. KAOS is a business goal-based technique for information system requirements engineering that addresses numerous requirements engineering features which include: system requirements elicitations, requirements analysis and requirements management. This technique also helps in the victorious completion of business requirements and helps to identify conflicts among IS and business sectors [30].

All these techniques have several drawbacks. First, the techniques are complex for information system analysts and developers to understand; also they provide lack of information on developing suitable information system for government sectors. Secondly, they do not offer sufficient information on business processes, as one business goal is a mixture of different sub-goals that need to be discovered in order to analyse the business goal entirely. Thirdly, the techniques are time consuming. In the latter part of the last century, with the quick increase in globalization in government and business sectors both need to move quicker and require quicker information system implementation, and have a strong association involving clear communication with information system.

## 3 Methodology: Business Process Modelling with UML

Recent studies on alignment between business and eGovernment research suggest two main IS theories in the context of internal and external organizational issues, namely system theory and network theory [29]. System theory is the interdisciplinary theory of IS in general, with the aim of identifying system requirements patterns and clarifying principles that can be distinguished from, and functional for all types of IS at all nesting levels in all fields of IS research. This theory's main focus is on internal organisational relationships [29]. Network theory examines the business organizational structure in relation to the organizational social aspect. These theories tend to place more emphasis on the business structures and dynamics of social relationships.

However, due to the nature of this research, we use system theory to underpin this study. A system is a combination of subsystems or parts that are integrated to accomplish the overall business goals [25, 29]. Moreover, the system in an organization has numerous inputs, which go through several processes to obtain accurate outputs to achieve the overall system goals. An organization is made up of many departments or divisions, such as finance, marketing, administration etc. and each department requires system support to achieve common business goals. However, the nature of the business often changes which distributes alignment processes between business departments and IS [29]. Systems theory helps in the understanding of the different organizational factors that contribute towards strong alignment and helps to identify the system requirements from the business process.

Moreover, in this paper we have chosen the modelling method is because it helps to visualize the business process. Unified Modelling Language (UML) is a language is being used in an industry widely. It has wide-range of elements and diagrams and it is a rich graphical notation. In this paper, UML has been used to visualize and to identify the business goals and Microsoft Visio tool has been used to model the business process.

# 4 Models for Alignment in EGovernment

Information and communication technologies (ICT) and IS are increasingly playing a vital role in the success of any type of business organization in the information age. The impact of IS on numerous domains, such as education, medicine, business, engineering, government etc. [23, 24] has been enormous. However, rapid change in both ICT and IS fields and substantial enhancements in digital connectivity has forced governments to change their structure. This is why mostly governments these days are reconsidering the way they work and interact both internally and with external business organizations [26, 28].

Moreover, this rapid change in ICT and IS has also encouraged government organizations and associations to reassess their internal and external relationships and dealings [4, 20]. Therefore, alignment between IS and other government agencies is important to establish strong working relationships. One way of aligning IS with other agencies in the government is to develop a system in accordance with the government's needs. This is only possible if IS developers understand and analyze the government environment, activities and business processes, all of which must be considered before commencing the development phase of IS. This is why business process and goal modelling is required before IS implementation to ensure the provision of appropriate government services [21]. In this paper, we only deal with the derivation of IS requirements from the eGovernment process, which is level 2 of the method, therefore, this paper will not discuss level 1 of the model.

An e-health process in the context of eGovernment has been selected to evaluate this proposed method. As the e-health process in the eGovernment infrastructure is the entity that focuses on the government's objectives in relationships to provide better health care services, modelling the e-health process is a key step towards obtaining IS requirements from the health process as it shows how the government's goals in relation to e-health can be attained using a real scenario [14, 24]. To model the e-health process, the well accepted standard OMG, namely Business Process Modelling Notation (BPMN) [29, 30] has been used. The technique is widely used in the area of system analysis and development. It is now considered to be a standard modelling language that bridges the gap between the development process and the business model.

### 4.1   Modelling eGovernment Process

The model used in this paper aims to assist eGovernment by helping e-health system developers better understand the eGovernment environment and to derive the system requirements from the eGovernment process and goals. The proposed model requires a clear understanding of e-health processes in regard to modelling e-health processes/goals and extracting the related goals effectively to generate e-health IS requirements. The model is categorized into two levels, as demonstrated in Fig. 1.

The first level of the proposed model is derived from the existing literature, which is based on an understanding of eGovernment infrastructure [20, 23]. This infrastructure is divided into four phases, namely: security and privacy, customer relationship, eGovernment services and business processes. This level demonstrates the decision level of an eGovernment infrastructure and details the government's objectives, the government services and resources and the eGovernment strategies and processes in the form of a vision, goal, evaluation and targets of the government strategy.

Level 2 of the model defines the concept of modelling eGovernment processes to obtain, model and investigate e-health processes in the context of eGovernment and the development of an e-health system. This level explains how to attain IS requirements from the health processes that were obtained through modelling the eGovernment processes based on a thorough understanding of the healthcare environment. This level is divided into four phases: extracting business goals, modelling goals, goals analysis and derivation of IS requirements.

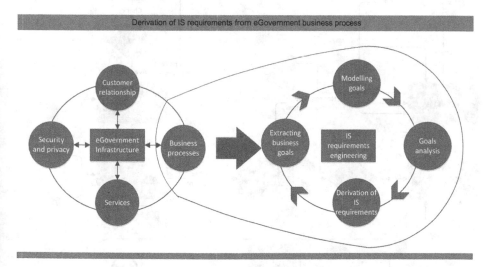

**Fig. 1.** Business processes within eGovernment

## 4.2   Modelling e-Health Process

Patient visits to a healthcare provider were used as the process to be modelled and to validate the proposed method, as shown in Fig. 2. This process was implemented by the Ministry of eGovernment in Saudi Arabia, with the aim of improving the sharing of

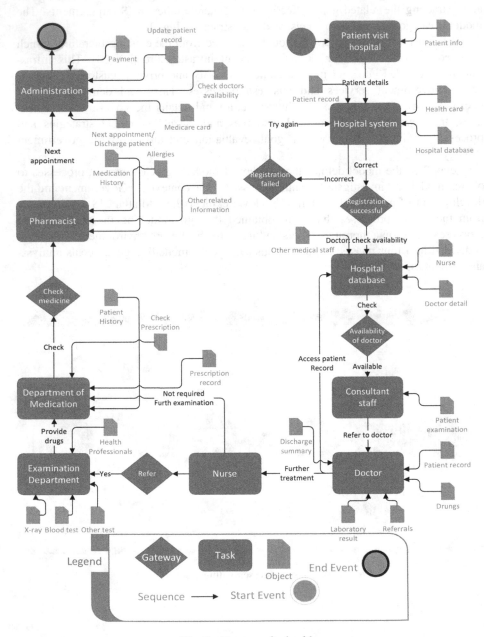

**Fig. 2.** Process of e-health

medical information among healthcare providers and the government so that e-health services could be improved in the kingdom.

As shown in Fig. 2, the selected e-health process is divided into four stages: patient registration and doctor allocation, providing consultation to the patient, further patient examination and discharge. In the first stage, the receptionist registers the new patient or retrieves the patient's data from the e-health system if the patient is registered already. In the second stage, the doctor commences the patient examination and stores the data related to the patient in the system, including laboratory results, patient medical history, referrals and prescriptions. This confirms that the doctor has examined the patient and understands the patient's condition.

In the third stage, if the patient requires further examination, such as a blood test, an X-ray, the doctor refers them to other medical departments or healthcare professionals. In the fourth stage, the hospital administrator organizes a follow-up patient visit if required, finalizes the payment and discharges the patient. The eGovernment process in the context of e-health as shown in Fig. 2 illustrates how the overall health infrastructure will be improved by allowing healthcare professionals to share information and provide access to medical records. Moreover, it demonstrates how medical practices may be improved once the e-health IS is adopted and integrated with other existing eGovernment IS.

## 4.3   Business Goals Modelling and Goals Analysis

The e-Government infrastructure includes many hardware devices, software applications and networks, and is utilized by many participants with different capabilities and technical skills. In terms of the e-health process and extracting business goals from the process, modelling and analysing the goals, these different e-health stakeholders and variety of components reduces e-health IS activities that are complex [17, 18]. Hence, the method of modelling and analysing eGovernment goals must be reliable enough to ensure the effective administration of the rapidly changing eGovernment processes and goals [6, 19].

Accurately modelling the eGovernment process is the key element in deriving the IS requirements. Modelling the business process shows how different government requirements can be executed through every process to achieve the proposed goals [13, 16]. A single eGovernment goal can carry many sub-goals that need to be identified for complete business process modelling and to accurately extract the IS requirements [14]. In this proposed method, once the health process in the context of the eGovernment environment has been identified and modelled, we then classify the sub-goals in the selected health process using the concept of the UML goal tree.

The goal tree in Fig. 3 represents the sub-goals, which are extracted from the model in Fig. 2. This goal tree model represents the number of eGovernment goals in the context of e-health and their related tasks. For example, goal "hospital" has two main sub-goals: "Hospital administration" and "Hospital system". The health goal "Hospital system" further sub-divided into nine sub-goals, namely, "Patient registration", "Health system integration", "Consultant", "Medical staff", "Consultation", "Medication",

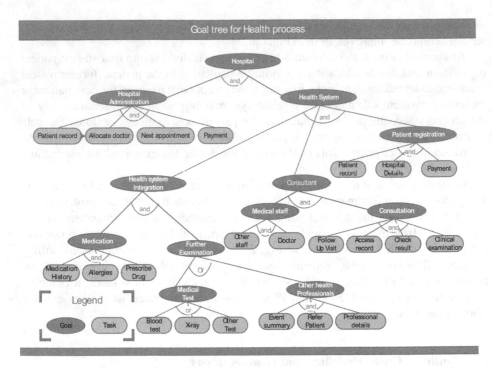

**Fig. 3.** Identification of goals and tasks in the e-health process

"Further examination", "Medical test" and "other health professionals". Moreover, this model prioritizes the goals, which helps to translate them into IS requirements.

Once the goals have been prioritized, we then analyze the eGovernment goals by addressing the following questions: who are the IS stakeholders, where in the eGovernment sector is the system going to be used; and why does eGovernment need this system? At this stage, the IS analysts inspect the model and label the goals and tasks to identify which ones are not important to develop from the government's perspective.

These goals and tasks are then marked with a cross and removed from the IS requirements. Figure 4 shows the analysis of the process, including the goals and their tasks. In this model, the health goals and tasks that cannot be developed technically have been removed from the model. Health goals and tasks that cannot proceed are as follows: Health goals "Medical Test" and Health tasks: "Other staff", "Clinical examination", "Prescribe drug", "Blood test", "x-ray", and "Other test".

### 4.4 Derivation of IS Requirements for eGovernment in the Context of e-Health

Once the business goals have been extracted from the proposed e-health process and analysed, the UML use cases are obtained from the UML goal tree in Fig. 4.

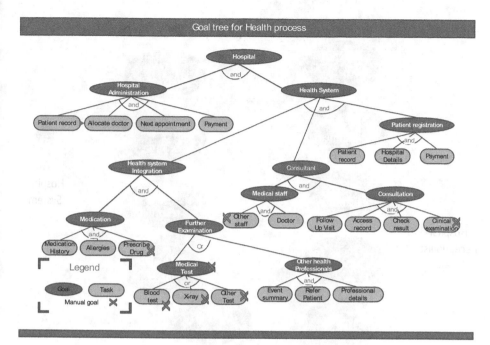

**Fig. 4.** Analysis of e-health goals and tasks

Figure 5 presents the UML use cases to show how eGovernment developers can develop an e-health system in the context of eGovernment. Use cases in this model demonstrate the complete IS requirements, which are categorized into four actors: the hospital receptionist, the hospital system, the doctor and the nurse. The hospital receptionist registers the patient's record and refers the patient to an available doctor. The doctor and nurse are the medical staff who examine the patient. The hospital system actor represents the database of the hospital, where the patient's records are stored.

The UML use cases allow the IS analysts to change or modify the IS requirements at any stage of the system development life cycle in order to remove ambiguity between the system stakeholders. IS analysts collect the details of the use case package and forward these to the developers to complete the system. The e-health developers in the eGovernment infrastructure first check the use case package to see whether the package has any duplicate records or requirements. If there are no errors, e-health IS requirements are ready to implement. At this level, the e-health IS requirements are clear and understood, which helps the developers to develop the system according to the government expectations. This ensures there is strong alignment between IS and other agencies in the eGovernment sector.

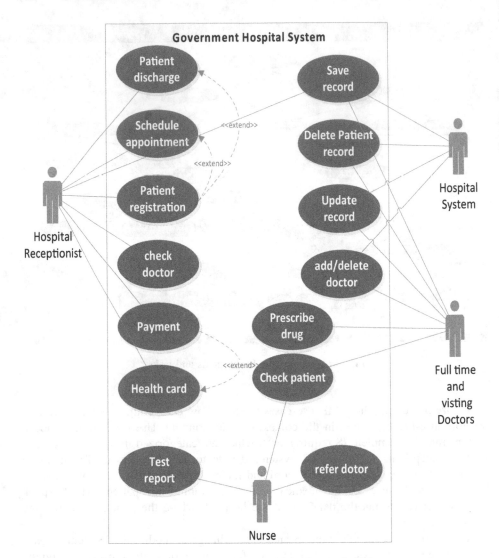

**Fig. 5.** e-Health system requirements in context of eGovernment

## 5  Conclusion and Implications

The impact of IS on numerous domains, such as education, medicine, business, engineering, governments has been enormous etc. However, rapid change in both ICT and IS fields and substantial enhancements to digital connectivity has forced governments to change their structure [4, 20]. This is why most governments these days are reconsidering the way they work and interact, both internally and with external business organizations. This rapid change in ICT and IS has also encouraged government organizations and associations to reassess their internal and external relationships and dealings. Therefore, alignment between IS and other government agencies is important to establish strong working relationships.

Alignment is the process where IS and other government agencies work together to reach a common government goal. The successful process of alignment helps governments to achieve the following: effective strategic planning towards better IS support to the government, better business performance for the government, bridging the communication gap at all levels of the government, and stronger relationships between IS and other government agencies [6, 8, 9, 18].

This paper provides an alignment solution to the eGovernment sector in the context of modelling business processes and developing suitable IS. One way of alignment business with information system the development of system according to the business expectation. The paper presents an alignment model, in which we model the eHealth business process, extract and analyses system requirements. This alternately allows IS developers to develop a system that meet the needs of eGovernment effectively.

Two main implications can be derived from this paper. First, it focuses the attention of eGovernment and alignment researchers on business process modelling in order to ensure more and accurate support from IS. This proposed method shows how modelling the business environment can lead to a complete understanding of the goals and values of eGovernment in order to enhance any business process in any sector. Second, generating IS requirements from the eGovernment process is a difficult task due to the complexity of the eGovernment infrastructure, as one business process carries more than one sub-goal. This method defines how the eGovernment business process can be measured and the related business goals can be extracted.

The proposed model has one limitation. The model is validated using only one business process in e-health, therefore, it is important to evaluate and validate the proposed model with different eGovernment business processes. As one business process vary from eGovernment aim to aim and from Government sector to sector. Thus, further investigation is required in validating the model with more than one eGovernment business processes in order to enhance the effectiveness of the model.

# References

1. Almajali, D., Dahlin, Z.: IT-business strategic alignment in influencing sustainable competitive advantage in jordan. In: Proceedings of the Annual International Conference on Infocomm Technologies in Competitive Strategies (2010)
2. Asato, R., Spinola, M.D.M., Costa, I., Silva, W.H.D.F.: Alignment between the business strategy and the software processes improvement: A roadmap for the implementation. In: 2009 Portland International Conference on Management of Engineering & Technology, PICMET 2009 (2009)
3. Bergeron, F., Raymond, L., Rivard, S.: Ideal patterns of strategic alignment and business performance. Inf. Manag. 41, 1003–1020 (2004)
4. Berghout, E., Tan, C.-W.: Understanding the impact of business cases on IT investment decisions: An analysis of municipal e-government projects. Inf. Manag. 50(7), 489–506 (2013)

5. Bharadwaj, A., El Sawy, O.A., Pavlou, P.A., Venkatraman, N.V.: Digital business strategy: Toward a next generation of insights. MIS Q. **37**(2), 471–482 (2013)
6. Bleistein, S.J., Cox, K., Verner, J.: Validating strategic alignment of organizational IT requirements using goal modeling and problem diagrams. J. Syst. Softw. **79**(3), 362–378 (2006)
7. Bubenko, J., Persson, A., Stirna, J.: EKD User Guide. Department of Computer (2001)
8. Charoensuk, S., Wongsurawat, W., Khang, D.B.: Business-IT Alignment: A practical research approach. J. High Technol. Manag. Res. **25**(2), 132–147 (2014)
9. Campbell, B.: Alignment: Resolving ambiguity within bounded choices. In: PACIS, vol. 54 (2005)
10. Chen, R.S., Sun, C.M., Helms, M.M., Jih, W.J.K.: Aligning information technology and business strategy with a dynamic capabilities perspective: A longitudinal study of a Taiwanese Semiconductor Company. Int. J. Inf. Manage. **28**(5), 366–378 (2008)
11. De Vara González, J.L., Díaz, J.S.: Business process-driven requirements engineering: A goal-based approach (2007)
12. Foss, N.J., Lindenberg, S.: Microfoundations for strategy: A goal-framing perspective on the drivers of value creation. Acad. Manag. Perspect. **27**(2), 85–102 (2013)
13. Gartlan, J., Shanks, G.: The alignment of business and information technology strategy in Australia. Australas. J. Inf. Syst. 14(2) (2007)
14. Jorfi, S., Jorfi, H.: Strategic operations management: Investigating the factors impacting IT-business strategic alignment. Procedia-Soc. Behav. Sci. **24**, 1606–1614 (2011)
15. Jaskiewicz, P., Klein, S.B.: The impact of goal alignment and board composition on board size in family businesses. J. Bus. Res. **60**(10), 1080–1089 (2007)
16. De la Vara González, J.L., Diaz, J.S.: Business process-driven requirements engineering: A goal-based approach. In: Proceedings of the 8th Workshop on Business Process Modeling (2007)
17. Lehtola, L., Kauppinen, M., Kujala, S.: Requirements prioritization challenges in practice. In: Bomarius, F., Iida, H. (eds.) PROFES 2004. LNCS, vol. 3009, pp. 497–508. Springer, Heidelberg (2004). doi:10.1007/978-3-540-24659-6_36
18. Martinez-Simarro, D., Devece, C., Llopis-Albert, C.: How information systems strategy moderates the relationship between business strategy and performance. J. Bus. Res. **68**(7), 1592–1594 (2015)
19. Meijer, A., Thaens, M.: Alignment 2.0: Strategic use of new internet technologies in government. Gov. Inf. Q. **27**(2), 113–121 (2010)
20. Mirchandani, D.A., Lederer, A.L.: The impact of core and infrastructure business activities on information systems planning and effectiveness. Int. J. Inf. Manag. **34**(5), 622–633 (2014)
21. Odiit, M.C.A., Mayoka, G.K., Rwashana, A.S., Ochara, N.M.: Alignment of information systems to strategy in the health sector using a systems dynamics approach. In: Proceedings of the Southern African Institute for Computer Scientist and Information Technologists Annual Conference 2014 on SAICSIT 2014 Empowered by Technology, p. 38 (2014)
22. Raup-Kounovsky, A., Canestraro, D.S., Pardo, T.A., Hrdinová, J.: IT governance to fit your context: two US case studies. In: Proceedings of the 4th International Conference on Theory and Practice of Electronic Governance, pp. 211–215(2010)
23. Ryu, H.S., Lee, J.N., Choi, B.: Alignment between service innovation strategy and business strategy and its effect on firm performance: an empirical investigation. IEEE Trans. Eng. Manag. **62**(1), 100–113 (2015)
24. Ullah, A., Lai, R.: A requirements engineering approach to improving IT-Business alignment. In: Pokorny, J., et al. (eds.) Information Systems Development, pp. 771–779. Springer, New York (2011)

25. Ullah, A., Lai, R.: A systematic review of business and information technology alignment. ACM Trans. Manag. Inf. Syst. (TMIS) **4**(1), 4 (2013)
26. Veres, C., Sampson, J., Bleistein, S.J., Cox, K., Verner, J.: Using semantic technologies to enhance a requirements engineering approach for alignment of IT with business strategy. In: 2009 International Conference on Complex, Intelligent and Software Intensive Systems, CISIS 2009, pp. 469–474 (2009)
27. White, S.A., Miers, D.: BPMN modelling and reference guide: Understanding and using BPMN. Future Strategies Inc., New York (2008)
28. Zowghi, D., Jin, Z.: A framework for the elicitation and analysis of information technology service requirements and their alignment with enterprise business goals. In: 34th Annual Computer Software and Applications Conference Workshops (COMPSACW), pp. 269–272. IEEE (2010)
29. Rakgoale, M.A., Mentz, J.C.: Proposing a measurement model to determine enterprise architecture success as a feasible mechanism to align business and IT. In: International Conference on Enterprise Systems (ES), pp. 214–224 (2015)
30. Dardenne, R., Lamsweerde, A., Fickas, S.V.: Goal-directed requirements acquisition. Sci. Comput. Program. **20**(1–2), 3–50 (1993)
31. Gordijn, J., Yu, E., Raadt, B.V.D.: e-Service design using I* and e3 value modeling. IEEE Softw. J. **23**(3), 26–33 (2006)
32. Ellis, C.A., Wainer, J.: Goal-based models of collaboration. Collaborative Comput. **1**(1), 61–86 (1994)
33. Card, S.K., Moran, T.P., Newell, A.: The Psychology of Human-Computer Interaction. Erlbaum, Hillsdale (1983)

# Visualization Perspectives

# Visualization for Decision Support in FITradeoff Method: Exploring Its Evaluation with Cognitive Neuroscience

Adiel Teixeira de Almeida[(✉)] and Lucia Reis Peixoto Roselli

Center for Decision Systems and Information Development – CDSID,
Federal University of Pernambuco – UFPE, Cx. Postal 7462, Recife, PE 50.630-970, Brazil
almeida@cdsid.org.br
http://www.cdsid.org.br

**Abstract.** FITradeoff method uses a flexible and interactive approach for supporting decisions in multicriteria problems in the context of MAVT (Multi-Attribute Value Theory) with partial information. Since the very beginning of the preference elicitation process, a subset of potential optimal alternatives (POA) is selected based on the current partial information provided. Then, the Decision Maker (DM) has the flexibility of interrupting the elicitation process for analyzing the partial result by other means, such as graphical visualization of performance of POA. This flexibility is available in the whole process. Evaluating the visualization confidence for decision support in FITradeoff method is crucial. Furthermore, information for designing of this visualization is relevant. This paper shows how these issues could be approached based on cognitive neuroscience, with particular focus given on eye tracking resources.

**Keywords:** Evaluating visualization · Graphical visualization · FITradeoff MCDM method · Cognitive neuroscience · Neuroeconomics · Decision neuroscience · Eye tracking

## 1 Introduction

The process of building a multicriteria decision model encompasses several relevant issues. One of the most important of these issues is that of evaluating the criteria (or attributes) weights [1]. This is particularly relevant when the aggregation procedure is based on an additive model, given the meaning of these weights in the context of MAVT (Multi-Attribute Value Theory) [2].

The elicitation of the criteria weights may be considered as a central issue in MCDM/A (Multi-Criteria Decision Making/Aid) process, as mentioned in a recent survey [3]. There many studies related to this matter and broad overview on the elicitation procedures of weights for additive models can be seen in [4].

Previous behavioral studies on the main elicitation procedures in the context of MAVT, considering aggregation of criteria with additive models have shown several issues regarding inconsistencies in the process of preference modeling [5, 6]. Possibly due to these issues, several MCDM/A methods for dealing with additive models have been proposed in the literature [1]. One of these methods is the FITradeoff (Flexible and

© Springer International Publishing AG 2017
I. Linden et al. (Eds.): ICDSST 2017, LNBIP 282, pp. 61–73, 2017.
DOI: 10.1007/978-3-319-57487-5_5

Interactive Tradeoff), which may reduce inconsistencies in the elicitation process [1], when compared with the traditional tradeoff procedure [2].

One of the main features of the FITradeoff method is the use of a flexible elicitation procedure [7] in order to reduce the timing spent in the elicitation process for establishment of criteria weights [1].

One of the mechanisms for introducing this flexibility in FITradeoff is the use of graphical visualization for analysis of intermediate results in the decision process.

In the preference elicitation process, FITradeoff selects a subset of potential optimal alternatives (POA), based on the current partial information available. The flexibility of FITradeoff allows the Decision Maker (DM) the possibility of interrupting the elicitation process in order to analyzing the partial results (POA) at each step in the process. This analysis may be conducted based on graphical visualization of performance of POA.

This process of visualization analysis is emphasized in this paper and illustrated with an application of FITradeoff for selecting a weapon to be incorporated in a navy ship [8]. It is shown that the visualization analysis may abbreviate the decision process, reducing the time spent by the DM. It is also shown that the decision could be based on the visualization information provided.

Therefore, evaluating the confidence of this visualization analysis is an important matter. Also, finding means for improving this confidence is relevant when preparing analysts for supporting DMs. Furthermore, information for the designing of this visualization is relevant in the process of building DSS (Decision Support Systems). This paper shows how these issues can be approached based on cognitive neuroscience, particularly when using eye tracking resources.

## 2   Application of FITradeoff DSS with Visualization

In this section the FITradeoff method [1] is briefly described followed by an application for selecting a weapon to be incorporated in a navy ship [8]. This paper shows how the use of the graph visualization can be improved in the method described in reference [1].

The decision process is considered with an emphasis to the use of graphical visualization in the analysis phase. The graph visualization for the elicitation questions is not considered in this paper, since it involves other issues to be analyzed.

### 2.1   FITradeoff Method

The FITradeoff method supports decisions in multicriteria problems in the context of MAVT, with partial information. This method uses a flexible and interactive approach [1] in such way that graphical visualization could be applied in order to shorten the modeling process.

Figure 1 shows the decision process with FITradeoff, in which preference statements are collected from the DM in step 1, in the preference elicitation process. These preference statements consist of choices the DM makes regarding a pair of consequences, which are well known for the traditional tradeoff procedure [2].

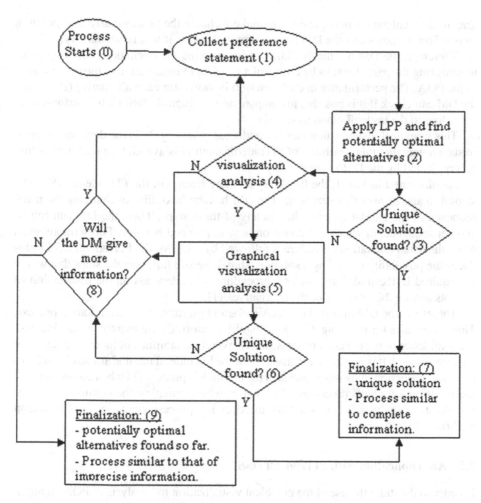

**Fig. 1.** FITradeof decision process.

Since the very beginning, for each input from the DM, a set of LPP (linear programming problems) are applied in step 2, in order to select a subset of potential optimal alternatives (POA). A more detailed explanation as to how the initial POA is obtained is given in [1]. This POA reflects the partial result with the new preference statement obtained and added to the current partial information.

If a unique solution is found (in step 3), then, the process is finalized (in step 7). Otherwise, the DM may conduct a graphical visualization analysis in steps 5 and 6. If the DM makes a decision of not conducting the visualization analysis, in step 4, and confirming to continue the elicitation process, in step 8, then a new preference statement is collected, returning to step 1.

The graphical visualization analysis in step 5 can be conducted at any cycle in the decision process and may result in a decision, in step 6, if the DM chooses one of the current POA as a solution for the decision problem. If the DM considers that the graphical visualization is not enough to make a choice the process carries on going to step 8. Step 9 shows that the DM can get a partial result, if that is satisfactory.

Therefore, the DM has the flexibility of interrupting the elicitation process in order to analyzing the partial results by means of a graphical visualization of the performance of the POA. The performance in each criterion is shown for each alternative (POA) and the DM can check if it is possible to compare and distinguish their global performance, according to the tradeoff amongst criteria.

This graphical visualization can be conducted at any cycle in the decision process, in such a way that this flexibility of FITradeoff method is available at each interaction accomplished by the DM in the process.

As illustrated in Fig. 1, the flexibility is a key feature in the FITradeoff DSS. The steps 4, 6 and 8 show this very well. The DM has the flexibility of changing the usual sequence of the elicitation procedure in any of these steps. Then, the DM can follow another sequence, in which a decision on how to proceed is given. Furthermore, even when the usual elicitation procedure is followed by the DM, the FITradeoff DSS introduces the possibility of finding a solution (step 3) before finishing the steps that would be required in the traditional tradeoff procedure. This depends on the distribution of weights and on the topology of the alternatives [1].

Therefore, the DM may make a decision before go further in the elicitation process. This contributes for reducing the DM's cognitive effort. Consequently, it may increase the confidence of the decision process, since several possibilities of inconsistencies and bias are avoided. It is worthy to emphasize that the FITradeoff method already introduces a marked reduction in inconsistencies in the elicitation process [1] when compared with the traditional tradeoff procedure [2]. That is why evaluating the confidence of this visualization analysis is important for the decision process supported by FITradeoff method.

## 2.2 An Application with FITradeoff DSS

In order to illustrate the use of the graphical visualization for analysis of POA an application of FITradeoff is given. This application is based on realistic cases and is related to selecting a weapon to be incorporated in a navy ship [8]. There are 30 alternatives to be analyzed and the following criteria are considered: Hit rate, Range, Fire rate, MTBF (mean time between failures), MTTR (mean time to repair), and Cost. All criteria are to be maximized, but MTTR and cost. This application has applied the FITradeoff DSS, which is available by request (www.fitradeoff.org).

Since the first phase of the elicitation procedure, in which the criteria weights are ranked, four alternatives are selected as POA. Then, the process continues and DM answer questions in the elicitation process in order to reduce the weight space and obtaining the solution to the problem. The questions asked to reduce the number of alternatives consist of the standard kind of question in the elicitation procedure, although in this case it is more specifically related to the scope of the traditional tradeoff procedure [2]. This is detailed in [1].

The number of questions will depend on the relative distribution of performance of alternatives and the distribution of weights [1]. The minimum number of questions in the traditional tradeoff procedure is expected to be $3(n - 1)$, where $n$ is the number of criteria [1]. In this case 15 questions would be expected. However, the problem is solved after the DM answer the fourth question.

The number of POA with questions answered by the DM is:

- after ranking the weights: 4 POA (alternatives: 20, 21, 23 and 28);
- after first question: 4 POA, same previous result;
- after second question: 3 POA (20, 23, 28);
- after third (till eighth) question: same result;
- after ninth question: 2 POA (23 and 28);
- after tenth question: 1 alternative - solution found to be alternative 28.

Therefore, the visualization analysis can be conducted between these questions and the DM may be able to choose one of the POA as a solution to the problem. This could reduce the timing for the decision process. That is, if the DM is able to make a decision with the graph visualization, no further question is required any more. It should be noted that, although this particular decision problem in the navy context has a reduced number of questions until a solution is found (see steps 3 and 7 of Fig. 1), it could take longer. Also, in general the DM has not an idea of how long this process would take. That is why estimations about reducing times cannot be provided. In the following section the visualization analysis for this problem is illustrated.

### 2.3  Visualization with FITradeoff DSS

The Graphical visualization of FITradeoff DSS considers several types of graphs that are available to the DM at any stage of the interactive process. These graphs are commonly applied in different multicriteria methods, which may have additional different types of graphs. For the purpose of this analysis, subsequently shown, these kinds of graphs are suitable for comparing alternatives and have been widely applied. Following Figures shows two types of graphs most applied in different stages of the decision problem previously described.

Figure 2 shows the bar graph in an early stage, when the ranking of weights is obtained. In this graph, the set of four POA are grouped by criterion. The criteria are shown in order of weights from the left to the right. The performances of alternatives are normalized in a ratio scale (0 to 1, the maximum performance).

In Fig. 2, it can be evaluated if the DM is able to compare the four alternatives with this graphical visualization. It might not be possible for a DM to analyze a set of information with four alternatives in six criteria.

As it can be seen, for the first criterion with the greatest weight, the four alternatives have almost the same performance. For the second criterion alternative 21 has a performance a bit above 0.6. For this second criterion the alternative 20 and 28 have the greatest performance (close to 1), followed by 23 (around 0.9). In the third criterion, the alternatives have diversified performance and with 23 has the best performance.

The DM could consider that at the very beginning of the process it is not worthy to make a decision with so many items to be considered (4 alternatives $\times$ 6 criteria).

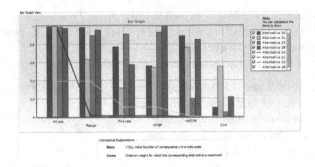

**Fig. 2.** Graphical visualization with FITradeof after weights ranking and first question - type bar graph.

Figure 3 shows another type of graphical visualization with similar information given in the Fig. 2. Same situation is observed in Fig. 3. The amount of information make not easy to the DM to make a decision.

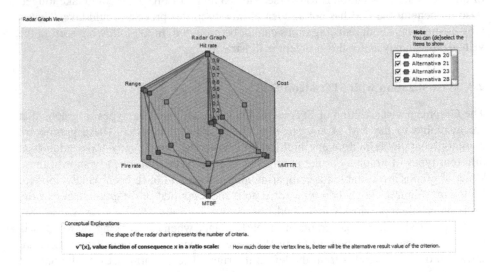

**Fig. 3.** Graphical visualization with FITradeof after weights ranking and first question - type radar graph.

Figure 4 shows the bar graph after the DM has answered the second question, when the set of POA is reduced to three alternatives.

In Fig. 4, compared with previous graph, it seems easier to the DM to be able to compare the three alternatives with this graphical visualization. The analyst should discuss with the DM about the confidence of this analysis and the timing to be spent in the elicitation process.

Figure 5 shows the bar graph after the DM has answered the ninth question, when only two alternatives are selected as POA.

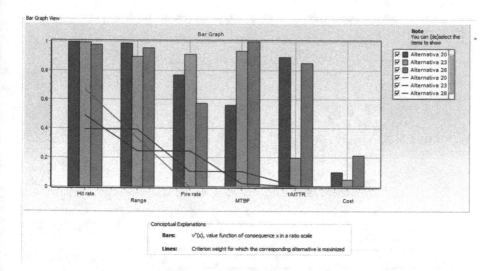

**Fig. 4.** Graphical visualization with FITradeof after second question - type bar graph.

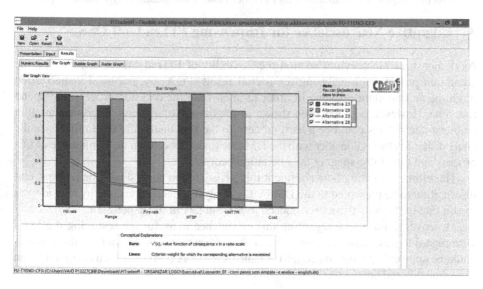

**Fig. 5.** Graphical visualization with FITradeof after third question - type bar graph.

Figure 6 shows the radar type of graphical visualization with similar information to that given in the Fig. 5.

Now, in both graphical visualizations, the DM may see that alternative 28 has a better performance compared to alternative 23 in four criteria. This means that, for this specific problem, a decision made at this point in the process might be reasonable. One cannot wonder if a decision with two POA would often be possible. However, it depends on how the performances of the alternatives are distributed by the criteria. This has to be evaluated case by case.

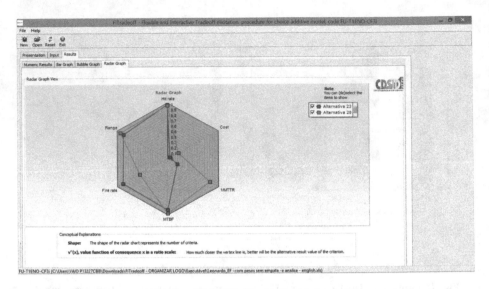

**Fig. 6.** Graphical visualization with FITradeof after third question - type radar graph.

# 3    Cognitive Neuroscience for Improving FITradeoff DSS

As demonstrated in previous section the flexibility of FITradeoff for interrupting the elicitation process and analyzing the partial result (POA) by graphical visualization might be an important resource in order to abbreviate the decision process and spend less time. However, as it has been shown in previous visualization, the differentiation amongst alternatives could be a hard process to the DM distinguishes. In that case the analyst should always consider with the DM, the tradeoff between the confidence of this process and time to be spent in the elicitation process.

Therefore, evaluating how confident this visualization analysis can be is important information to be presented to the DM by the analyst. This can be done by considering the hit rate of the best alternative chosen by the subject in the experiment.

Moreover, information regarding this confidence is useful for designing the DSS, with regard to this graphical visualization. In this section it is shown how these issues could be approached based on cognitive neuroscience. A particular focus is given on an experiment based on eye tracking resources.

## 3.1    Cognitive Neuroscience and Neuroeconomics

Cognitive neuroscience is concerned with cognition aspects and the biological processes, considering neural connections in the brain and its association with mental processes. Cognitive neuroscience is related to several fields and mainly to psychology and neuroscience [9]. It may have connections with many other areas, such as linguistics, philosophy, and mathematics. Experimental procedures from cognitive psychology are one of the methods employed in cognitive neuroscience that concerns this work. In association with cognitive neuroscience, neuroeconomics is closer related to the present study. It seeks to explain decision making in association with mental processes, considering how neuroscience findings are associated to decision models and economics.

Several different procedures and tools are used to study the biological basis of decision making and economic behavior. Analyzing which areas of the brain are most active while processing specific steps in the decision process may be useful to explain mental decision process and support the design of DSS. For that purpose neural imaging can be applied in human subjects, which can be done by fMRI [10] or EEG (Electroencephalography), for instance. These devices can give details of the brain with information on specific structures involved in a step of the decision process. The use of this kind of information and other related, such as response times, eye-tracking and neural signals that DM generate in decision making process, can be associated to the decision analysis process.

Associated to Cognitive neuroscience and neuroeconomics, other topics have been mentioned in the literature, which includes: Decision Neuroscience [11], Neuro-information system, Neuromanagement, and Neuromarketing.

Regarding the graphical visualization in FITradeoff the use of EEG may contribute to the investigation of areas of brain that are activated during the elicitation process, comparing for different types of visualization. It is important to evaluate which kind of visualization stimulates either an intuitive or a rational process, for instance. It can be explored the mechanism behind visualization decision-making, for elicitation and for analysis, examining how it interchanges with system 1 and system 2 [12] in the decision-making process.

In a recent study, eye-tracking experiment has been applied in order to identifying causal effects of intuition and deliberation [13]. One of the intended results is to identify how thinking mode affects the pattern for information collection in decision making.

Response times and fixation time in different regions of a graphical visualization are analyzed with eye-tracking in order to get insights for improvement of the design for the FITradeoff DSS.

Regarding response time one of the main features found to be analyzed in literature is the Hick's Law [14]. According to Hick's Law the time spent by the DM to make a choice among items is expected to linearly increase with the logarithm of the number of items.

In MCDM/A the items are not presented in the same way as it is in the previous neuroscience studies for Hick's Law. For those studies the choice is made by comparing single options, such as in a study [15] related to food choices. In that study they applied an eye-tracking experiment in which subjects have chosen snack food items.

In graphical visualization for the FITradeoff DSS the items are not presented as single choices. The items are presented in a combination of alternatives and criteria. That is, the alternatives to be analyzed are decomposed in criteria. It has to be investigated how the Hick's Law works in such a situation and how this can be applied to designing of DSS and recommendations for supporting analysts in their interaction with DMs.

There is another issue in graphical visualization for the FITradeoff. How should be considered the number of items, since there is a combination of alternatives times criteria? For instance, in Fig. 6 there are 4 alternatives and 6 criteria. One may wonder what would be the meaning of the number of items in Fig. 6. Would it be four alternatives? It could be considered 24 items (4 × 6)? Actually, it seems that the DM analyzes 4 items (alternatives), each one of them composed of 6 sub-items. Another possibility is that the DM analyzes how the alternatives perform for each criterion. For instance, which is the best alternative or the two best of them, for each criterion; or only for part of the criteria with higher weight?

## 3.2 Neuroscience Analysis for Visualization of FITradeoff DSS

An experiment has been design for exploring the use of cognitive neuroscience for evaluating the visualization for decision support in FITradeoff method. This experiment is following described with some preliminary results. So far, it has been observed an enormous potentiality with this kind of experiments in order to bring insights and new understanding on the decision process with visualization, which could be of interest for both: designing visualization in DSS and preparing the analyst for advising the DMs in a more proper way.

At this stage, the experiment considers visualization with bar graphs predominantly and including a few questions with other kind of graphical visualization.

The experiment has recorded data throughout the visualization analysis by the subjects, regarding eye position, duration in regions of interest and eye-movements. These recordings are done while performing different tasks. The basic tasks consists in choosing the best alternative from a set of alternatives, whose performance are shown for a set of criteria in graphs similar of those in Figs. 2, 3 and 4. In order to make their choices, the subjects have had as much time as they wished.

The set of alternatives are organized in such a way that they are ranked with a difference in performance of 5%, considering an additive model for aggregating the criteria.

Still, it has been thought about performing different tasks. For difference tasks it has been applied different number of alternatives × criteria, which included a combination of 3, 4 and 5 for both. This has made 9 different graphs to be analyzed. Another issue for differentiating the tasks is the distribution of weights of criteria, using a set of 9 graphs with equal weights and another set with skewed weights (decreasing form right to left order of visualization). It is worthy to note that these weights have been applied to design a set of alternatives, ranking them and computing their performance with 5% of difference.

## 3.3 Results of Eye Tracking Experiment

In the preliminary phase of the experiment several observations have been made and will be applied for designing other steps of this research. Nevertheless, some of these results are still being applied for designing changes in the DSS visualization and for instruction to the analysts regarding the use of visualization analysis in FITradeoff. Some of these observations are shortly described.

One of the information of this experiment regards the hit rate of the subjects, which can show how the confidence of graphical analysis changes with the number of items, for instance.

Concerning to the total number of items (alternatives × criteria), it has not been observed, so far, a correlation with the hit rate. This may confirm the conjecture on the association of timing and difficulty with the number of alternatives grouped by criteria or criteria grouped by alternatives. Similarly, Hick's Law does not seem to work in such a situation with the total number of items, although additional investigation is needed relating to this issue.

Regarding the two sets of graphs with different distribution of weights, following results have been found, from equal weights set to that of skewed weights: (a) the hit rate decreased in around 20%; (b) the mean time of fixation has increased of around 15%; (c) there is a high correlation between hit rate and scholar degree and degree of knowledge on MCDM/A methods.

Regarding the pupil size it has been noted a constriction of the pupil for the experiment of the set of graphs with skewed weights, when compared the set with equal weights.

Those subjects with lower hit rate have shown pupil with higher dilatation. As regard to the left and right pupillary response, it has been observed that there are not significant differences in variation all along the experiment. In general the size of left pupil is higher than the right one.

The experiment has considered regions of interest to be analyzed. The graph has been divided mainly in three regions: left, middle and right. A fourth region of interest related to x-axis has been investigated, but practically no interest has been demonstrated by the subjects.

There was not a significant difference of areas of interest in the graphs for the two types of set of graphs with different distribution of weights (equal and skewed weights). The middle region is the area of most interest by the subjects, when analyzing the graphs. It seems that for the first set of graphs with equal weights, there is not a common pattern, when the number of alternatives varies, although in most cases there is predominance for seeking the middle region of the graph.

The preliminary results are being applied for formulating more specific research questions in order to advance in this research. Also, some findings can still be considered for improving the graphical visualization in decision processes. For instance, the quality of data visualization and its use for decision making can be shown with the analysis of the hit rate of each combination of number of alternatives × criteria. This analysis can give basis to the analyst on how advising the DM in making decisions with particular number of items in a graph visualization.

A tradeoff has to be made in the use of visualization, considering two factors: (a) its use reduces the timing in the decision process and may contribute to its quality; (b) the confidence of a straight analysis of those graphs by the DM. The latter indicates that the confidence decreases with the amount of items in the graph. The former is a well know result in behavioral decision making [5, 6] the probability of inconsistences increases with the number of questions put forward to the DM.

## 4   Conclusions

This paper has focused on the process of visualization analysis of FITradeoff method, which is an important feature for its flexibility. In order to illustrate the use of graphical visualization in intermediate steps of analysis with FITradeoff, an application has been considered. As it has been shown, the visualization analysis may contribute to reduce the timing in the decision process, which may contribute to its quality. This is straightforward as it can be seen in Fig. 1, since a decision with the visualization (step 6) stops the elicitation process not requiring additional question (step 8) to the DM. In this sense this work deals with quality of data visualization and its use for decision making/support.

Therefore, evaluating the confidence of this visualization analysis is an important matter. Then, finding means for improving this confidence is relevant in the designing of this visualization within the process of building DSS. Some considerations are given on how to approach these matters, based on 'decision neuroscience' field, which is correlated to cognitive neuroscience, and neuroeconomics.

The study has been focused in a particular MCDM/A method based on a compensatory rationality by aggregating the criteria with the additive model. This method uses visualization as an important feature for analysis, incorporating more flexibility in the decision process, although these tools can be dismissed by following the interactive elicitation process.

For future work, the neuroscience analysis of graph visualization for the elicitation questions is going to be considered, since it contains other matters that require specific examination. Also, it can be considered a non-compensatory rationality [16], although an evaluation of the most appropriate kind of rationality (compensatory or non-compensatory), for each DM, remains to be made.

**Acknowledgments.** This study was partially sponsored by the Brazilian Research Council (CNPq) for which the authors are most grateful.

## References

1. de Almeida, A.T., de Almeida, J.A., Costa, A.P.C.S., de Almeida-Filho, A.T.: A new method for elicitation of criteria weights in additive models: flexible and interactive tradeoff. Eur. J. Oper. Res. **250**, 179–191 (2016)
2. Keeney, R.L., Raiffa, H.: Decision Making with Multiple Objectives, Preferences, and Value Tradeoffs. Wiley, New York (1976)
3. Riabacke, M., Danielson, M., Ekenberg, L.: State-of-the-art prescriptive criteria weight elicitation. Adv. Decis. Sci. (2012). ID 276584
4. Eisenführ, F., Weber, M., Langer, T.: Rational Decision Making. Springer, Heidelberg (2010)
5. Weber, M., Borcherding, K.: Behavioral influences on weight judgments in multiattribute decision making. Eur. J. Oper. Res. **67**, 1–12 (1993)
6. Borcherding, K., Eppel, T., von Winterfeldt, D.: Comparison of weighting judgments in multiattribute utility measurement. Manag. Sci. 37, 1603–1619 (1991)

7. de Almeida, A.T., Costa, A.P.C.S., de Almeida, J.A., Almeida-Filho, A.T.: A DSS for resolving evaluation of criteria by interactive flexible elicitation procedure. In: Dargam, F., Hernández, J.E., Zaraté, P., Liu, S., Ribeiro, R., Delibasic, B., Papathanasiou, J. (eds.) EWG-DSS 2013. LNBIP, vol. 184, pp. 157–166. Springer, Cham (2014)
8. de Almeida-Filho, A.T., Pessoa, L.A., Ferreira, R., de Almeida, A.T.: A navy weapon selection throughout fitradeoff. INFORMS Nashville, p. 65 (2016)
9. Kosslyn, S.M., Andersen, R.A.: Frontiers in Cognitive Neuroscience. MIT Press, Cambridge (1992)
10. Paulus, M.P., Hozack, N., Zauscher, B., McDowell, J.E., Frank, L., Brown, G.G., Braff, D.L.: Prefrontal, parietal, and temporal cortex networks underlie decision-making in the presence of uncertainty. NeuroImage **13**, 91–100 (2001)
11. Smith, D.V., Huettel, S.A.: Decision neuroscience: neuroeconomics. WIREs. Cogn. Sci. **1**, 854–871 (2010)
12. Kahneman, D.: Thinking Fast & Slow. Farrar, Straus and Giroux, New York (2011)
13. Hausfeld, J., Posadzy, K.: Tracing intuition and deliberation in risky decision making for oneself and others. In: 14th Annual Meeting, Berlin (2016)
14. Hick, W.E.: On the rate of gain of information. Q. J. Exp. Psychol. **4**(1), 11–26 (1952)
15. Thomas, A., Krajbich, I.: Simple economic choice in large choice sets: an investigation of Hick's law. In: Society For Neuro economics (SNE), Annual Meeting (2016)
16. de Almeida, A.T., Cavalcante, C.A.V., Alencar, M.H., Ferreira, R.J.P., de Almeida-Filho, A.T., Garcez, T.V.: Multicriteria and Multiobjective Models for Risk, Reliability and Maintenance Decision Analysis. International Series in Operations Research & Management Science, vol. 231. Springer, New York (2015)

# Incorporating Uncertainty into Decision-Making: An Information Visualisation Approach

Mohammad Daradkeh[✉] and Bilal Abul-Huda

Department of Management Information Systems, Faculty of Information Technology and Computer Sciences, Yarmouk University, Irbid, Jordan
{mdaradkeh,abul-huda}@yu.edu.jo

**Abstract.** Incorporating uncertainty into the decision-making process and exposing its effects are crucial for making informed decisions and maximizing the benefits attained from such decisions. Yet, the explicit incorporation of uncertainty into decision-making poses significant cognitive challenges. The decision-maker could be overloaded, and thus may not effectively take the advantages of the uncertainty information. In this paper, we present an information visualisation approach, called RiDeViz, to facilitate the incorporation of uncertainty into decision-making. The main intention of RiDeViz is to enable the decision-maker to explore and analyse the uncertainty and its effects at different levels of detail. It is also intended to enable the decision-maker to explore cause and effect relationships and experiment with multiple "what-if" scenarios. We demonstrate the utility of RiDeViz through an application example of a financial decision-making scenario.

**Keywords:** Decision support · Information visualisation · Interaction design · Risk · Uncertainty · What-if analysis

## 1 Introduction

Decision-making is one of the basic cognitive processes of human beings by which a preferred alternative is chosen from among a set of available alternatives based on input information [1, 2]. Ubiquitous in realistic situations, the information on which decisions are based is subject to uncertainty arising from different sources. Common sources include the lack of knowledge of true values of decision variables and future possibilities and outcomes. For example, the decision about whether to invest in a new product depends on the uncertain market conditions (e.g. whether the demand will go up or down). The possible outcomes of the decision (e.g. making a profit or loss) are also dependent on how much the demand goes up or down and its interaction with other variables (e.g. the price of the product). In this situation, the decision-maker may evaluate the outcomes and their likelihoods under different scenarios, and base his or her decisions on this evaluation [3]. Such decisions are inherently risky as available alternatives will generally involve some chance of unacceptable outcomes.

Ignoring uncertainty may simplify the decision-making process, but it may result in ineffective and less informed decisions. Thus, the uncertainty should be addressed from

© Springer International Publishing AG 2017
I. Linden et al. (Eds.): ICDSST 2017, LNBIP 282, pp. 74–87, 2017.
DOI: 10.1007/978-3-319-57487-5_6

the beginning of the decision-making process as an integral part of the information on which decisions are based. However, the incorporation of uncertainty into decision-making poses significant cognitive challenges. It adds complexity and confusion to the task of decision-making which is already complicated. One example of such confusion occurs in the evaluation or ranking of multiple viable alternatives. The decision-maker could also be overloaded, and thus may not effectively take the advantages of the uncertainty information [4, 5]. Moreover, the process of incorporating uncertainty into decision-making is a highly technical subject, and often not transparent or easy to grasp by decision-makers who lack necessary numerical skills.

Information visualisation provides an effective means for amplifying and enhancing the cognitive abilities of decision-makers to process and use information in decision-making [6]. It can facilitate the interaction between the decision-maker and decision model by converting it from a difficult conceptual process into a simple perceptual process [7]. It also can help decision makers to explore cause and affect relationships and analyse multiple "what-if" scenarios. The ability to analyse "what-if" scenarios is a key requirement for developing understanding about the implications of uncertainty, which in turn leads to making better informed and justifiable decisions [8]. According to Tufte [9] "Assessments of change, dynamics, and cause and effect are at the heart of thinking and explanation. To understand is to know what cause provokes what effect, by what means, at what rate."

In this paper, we present an information visualisation approach, called RiDeViz, to facilitating the incorporation of uncertainty into the decision-making process and making the decision-maker aware of its effects. To demonstrate the utility and practical use of RiDeViz, we apply it to an application example of a financial decision-making scenario.

## 2    Related Work

Numerous approaches have been developed for integrating uncertainty into information for visualisation, particularly by the geographical and scientific visualisation communities. Pang et al. [10] broadly classified the uncertainty visualisation approaches into adding glyphs, adding geometry, modifying geometry, modifying attributes, animation, sonification, and psycho-visual approaches. Davis and Keller [11] asserted that the use of colour value, colour hue, and texture are the best potential choices for representing uncertainty on static maps. However, how well these approaches provide decision support still needs further research.

Amar and Stasko [12] proposed a set of rationale precepts for integrating uncertainty into visualisation tools to support decision-making under uncertainty. These are:

(1) Expose uncertainty in data measures and aggregations, and show possible effects of this uncertainty on outcomes;
(2) Clearly present what comprises the representation of a relationship, and present concrete outcomes where appropriate; and
(3) Clarify possible sources of causation.

Some traditional approaches for conveying statistical and bounded uncertainty have also been proposed [13, 14]. These include error bars, box plots, range plots, scatter plots, histograms, and distributions. These approaches typically encode the minimum, maximum, mean, and quartile information of a distribution. While this information may be enough to convey the lack of certainty, these approaches focus on presentation rather than interactive analysis and exploration of uncertainty and its potential effects. Without the interactive exploration of uncertainty, the decision-maker may not fully appreciate the impact of uncertainty in the decision problem. According to the NIH/NSF visualisation research challenges report [15], a visualisation system to support decision-making should allow ordinary people to assess changes, cause and effects, and experiment with multiple "what-if" scenarios.

Our approach presented in this paper focuses not only on the presentation of uncertainty, but also on its analysis and exploration at different granularities of detail. In addition, it proposes translating the uncertainty information into a measure that can be used intuitively and easily in decision-making, as discussed in the next section.

## 3   Risk-Based Approach for Incorporating Uncertainty into Decision-Making

If uncertainty is incorporated into decision-making, the criteria used to assess the decision alternatives should reflect this. It is well-recognized that, in the presence of uncertainty, the risk of obtaining unacceptable outcomes is a frequently used criterion for exposing the effect of uncertainty and evaluating the decision alternatives [16]. This is because the risk of unacceptable outcomes offers a clear way to make sense of uncertainty and address it explicitly in the decision-making process [17].

Our approach for incorporating uncertainty into decision-making is to convert the whole process into one of determining the risk associated with the decision. This approach is shown in Fig. 1 where the decision-maker specify the risk criterion to be used and also the uncertainty in input information. For example, in the case of considering an investment decision problem, the two component of the risk might be the probability of making a loss and the amount of money that could be lost as a consequence of making a decision. The decision-maker is then interested in both the risk that the investment will make a loss, and how that risk is affected by his or her knowledge of the uncertainty in the information related to this particular investment.

In the proposed approach, the uncertainty in the input variables is represented as a range of possible values. For example, in an investment decision problem, if the decision-maker is uncertain about the exact cost of an investment, he or she can specify a range of values that reflects the uncertainty in the cost of that investment. There are many different methods of representing uncertainty such as Bayesian statistics, Dempster-Shafer theory, imprecise probability theory, random set theory, rough set theory, and fuzzy set theory [18]. However, these methods are mathematical in nature; hence, it is difficult for decision-makers to utilize them for specifying the amount of uncertainty in the input variables. In contrast, representing uncertainty as a range of possible values

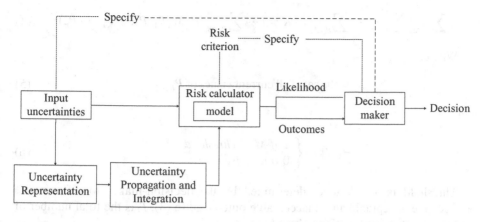

**Fig. 1.** The proposed approach for incorporating uncertainty into decision-making.

is intuitive and easy to learn, particularly for decision makers who do not have the technical expertise to fully understand the statistical results [19].

Once the uncertainties in the input variables are represented, they are then propagated through the model and integrated in such a way as to yield a risk distribution (i.e. range of possible outcomes and their likelihoods). For each risk calculation, it takes a value from an input variable and every possible value of the other input variables and calculates what proportion of these values will result in an unacceptable outcome. This process is repeated until completing all possible combinations, and the resulting range of risk values is tallied in a form of distribution. The distribution of risk values and the uncertainties in the input variables are discretized into a finite number of divisions to make it easier for the decision-maker to explore the relationship between the uncertainty and the resulting risk of unacceptable outcomes.

To put some structure, these steps are expressed in the following formulas:

Assume we have a decision model $M$ which has $N$ parameters $P$. Then, for any given set of parameter values, $M(P)$ will give a value for the model. Each parameter $P_i$ can take on $K_j$ discrete values $P_{ij}\{1 \leq j \leq K_j\}$ such that:

$$\text{Each } P_{i,j+1} = P_{i,j} + \Delta P_i \tag{1}$$

$$\text{Where } \Delta P_i = \left(P_{i,k} - P_{i,1}\right)/K \tag{2}$$

Note that $K_j$ may vary as the user selects and deselects values. Then, the probability of an unacceptable outcome $Pr_{i,j}$ if $P_i$ has a particular value $P_{i,j}$ is given by:

$$Pr_{i,j} = \frac{C_{i,j}}{T_i} \tag{3}$$

Where $C_{i,j}$ is the count of the number of parameter combinations that produce an unacceptable outcome when $P_i$ has a particular value $P_{i,j}$ given by:

$$C_{i,j} = \sum_{P_1 \in S_1} \sum_{P_2 \in S_2} \cdots \sum_{P_k \in S_k} \cdots K \neq i \cdots \sum_{P_N \in S_N} H\left[M\left(P_1, P_2, \cdots P_{i,j}, \cdots P_n\right)\right] \qquad (4)$$

Where:

$$S_i = \text{the set of all current values of } P_i \qquad (5)$$

and

$$H[M] = \begin{cases} 1 & \text{if } M < \text{Threshold} \\ 0 & \text{otherwise} \end{cases} \qquad (6)$$

Threshold is a risk level determined by the decision-maker to differentiate between the acceptable and unacceptable outcomes. In (3), $T_i$ is the total number of other parameter combinations given by:

$$T_i = \prod_{j=1 \cdots n, j \neq i} K_j \qquad (7)$$

Our approach to addressing uncertainty and its associated risk captures all possible outcomes; i.e., the acceptable and unacceptable ones as well as their relative probabilities. Thereby, it provides decision-makers with a complete risk profile for each alternative. The risk profile captures the essential information about how uncertainties in the input variables affect an alternative. It also provides a consistent basis for comparing these uncertainties; thereby allowing decision-makers to focus on key uncertainties that might significantly influence the consequences of their decisions.

## 4   Case Study: Financial Decision-Making Problem

To demonstrate the utility of our approach, we apply it on a case study of a financial decision-making problem concerning the selection of an investment based on uncertain information. Some examples of such a problem include the decision on whether or not to buy a property for investment or decision to select from among a set of projects available for investments. In making such decisions, decision-makers usually specify evaluation criteria (e.g. the potential profit and risk of making a loss associated with the investment). To predict and analyze the profitability of an investment, a financial model for investment decision-making called Net Present Value (NPV) is commonly used [20–23]. The purpose of NPV is basically to estimate the extent to which the profits of an investment exceed its costs [23]. A positive NPV indicates that the investment is profitable, while a negative NPV indicates that the investment is making a loss. A basic version of calculating NPV is given by Eq. 8:

$$NPV = C_0 + \sum_{t=1}^{n} \frac{(CI_t - CO_t)}{(1 + r)^t} \qquad (8)$$

Where

$C_0$    is the initial investment
n    is the total time of the investment
r    is the discount rate (the rate of return that could be earned on the investment)
$CI_t$    is the cash inflow at time t
$CO_t$    is the cash outflow at time t

As shown in Eq. 8, the NPV model consists of five input variables in its basic form. In practice, each of these variables is subject to uncertainty because the information available on their values is usually based on predictions, and fluctuations may occur in the future. Consequently, the investment decision can lead to many possible outcomes (i.e. different values of NPV). Since not all possible outcomes are equally acceptable to the decision-maker, the investment decision involves a degree of risk. The risk is present because there is a chance that the investment decision can lead to an unacceptable rather than an acceptable outcome.

## 5    Requirements and Design Considerations

To design effective visualisations for decision support, the designer must first understand what type of information is available about the decision problem, how information is processed by humans and how decisions are made in reality [24]. Based on the availability of information, decision-making problems can be classified into: (1) decision-making under certainty; (2) decision-making under risk; and (3) decision-making under uncertainty [25–27]. The decision problem under uncertainty and risk is usually characterised as consisting of the following main elements: (1) the set of alternatives from which a preferred alternative is chosen; (2) the uncertain input variables and their possible values; (3) the possible outcomes resulting from uncertainties in the input variables and their propagation through models and criteria used in decision-making; (4) the risk of obtaining undesirable outcomes associated with each alternative [28]. All these elements should be taken into consideration when designing visualisation tools to support informed decision-making. This is because in the presence of uncertainty and risk, decision-makers usually base their decisions not only on the possible outcomes but also on the uncertainty and risk each alternative entails.

In addition to all aforementioned information, decision-makers need to be able to explore and compare alternatives at different granularities of detail. The presence of uncertainty in the values of input variables implies that there are many possible realisations (or values) for each input variable. This gives rise to the presence of many possible scenarios, where each scenario represents a possible combination of all values of input variables, one for each variable [29]. The set of possible scenarios captures all possible outcomes and the range of uncertainties and risk anticipated. Kaplan and Garrick (1981) suggested that risk can be fully defined by a set of three components, namely: (1) a set of mutually exclusive and collectively exhaustive scenario conditions under which the possible outcomes may be realized, (2) a set of outcomes for each possible scenario, and (3) a probability of occurrence for each possible scenario.

The information visualisation tool should provide facilities for generating possible scenarios and conducing analysis based on the generated scenarios. This requires facilities for enabling decision-makers to provide their own estimates of the values for each uncertain variable and its distribution. In addition, it requires the provision of computational facilities for propagating all uncertainties through models and criteria used in decision-making. Once all uncertainties are propagated through models, the information visualisation tool should then provide decision-makers with a complete picture of the generated scenarios and the distribution of uncertainties and risks anticipated to exist in these scenarios. At the same time, it should allow decision-makers to interact with the decision model for experimenting with possible "what-if" scenarios and explore the outcomes and risk associated with alternatives under these scenarios.

## 6    Description of RiDeViz

The design of RiDeViz consists of two main parts: Decision Bars and Risk Explorer (see Fig. 2). Decision Bars, Fig. 2(a), provide overview information on the decision alternatives, the range of possible outcomes, and the risk of unacceptable outcomes associated with each alternative. Using these bars, the decision-maker can evaluate and then focus on specific alternatives for detailed analysis and exploration. In contrast, Risk Explorer, Fig. 2(b), provides a detailed view of the uncertainty and risk of unacceptable outcomes associated with each alternative. Using Risk Explorer, the decision-maker can interactively compare alternatives and analyse the risk of unacceptable outcomes associated with these alternatives at different levels of detail. In the following, we describe Decision Bars and Risk Explorer in a more detail.

### 6.1    Decision Bars

The Decision Bars interface shown in Fig. 2(a) consists of two panels: the Outcome Bars and Risk Bars. The Outcome Bars, shown in the top panel of Fig. 2(a), present the decision alternatives, each of which is visualized by a bar. For example, the Outcome Bars in Fig. 2(a) shows that there are ten investment alternatives for the decision problem at hand. The length of the bar represents the range of possible outcomes associated with the corresponding alternative. The black part of each Outcome Bar represents the mean value of possible outcomes. The dashed blue line along each bar represents the probability distribution of possible outcomes for the corresponding alternative.

The Outcome Bars enable decision-makers to identify the range of possible outcomes and the worst and best possible outcomes for each alternative. They also allow them to identify the relative likelihood of occurrence of the possible outcomes.

The Risk Bars shown in the bottom panel of Fig. 2(a) display the risk of unacceptable outcomes (in this case, the probability of negative NPVs that will make a loss). The height of the bar represents the degree of risk of unacceptable outcomes associated with the corresponding alternative. The higher the bar, the greater the risk of unacceptable outcomes. The decision-maker can use the risk information for evaluating, comparing

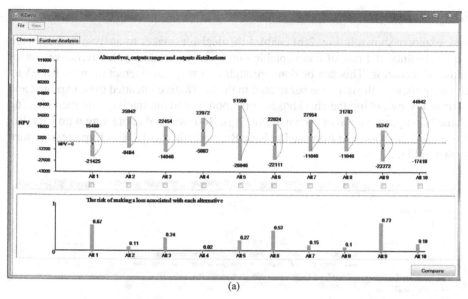

(a)

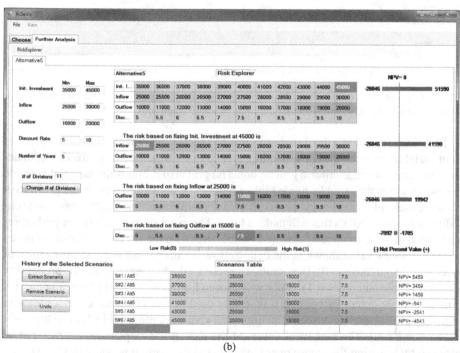

(b)

**Fig. 2.** Overview of RiDeViz: (a) Decision Bars and (b) Risk Explorer.

and then choosing preferred alternatives based on the level of risk he or she is willing to accept.

## 6.2   Risk Explorer

Risk explorer shown in Fig. 2(b) enables the decision-maker to analyse and compare the uncertainty and risk of unacceptable outcomes associated with alternatives under multiple scenarios. This can be done through two ways of interaction with RiDeViz: either by clicking the outcome bar related to the alternative intended to be explored and analyzed or by ticking the checkboxes corresponding to alternatives and then clicking on the "Compare" button as shown in Fig. 2(a). Both ways of interacting with RiDeViz move the decision-maker from the Decision Bars interface to the Risk Explorer interface shown in Fig. 3.

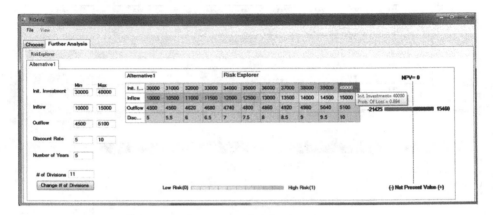

**Fig. 3.** A screenshot of Risk Explorer after selecting Alternative1 for analysis and exploration of risk of unacceptable outcomes under multiple "what-if" scenarios.

Risk Explorer allows the decision-maker to specify the range of possible values of input variables through the corresponding text boxes. Then, it portrays the risk of unacceptable outcomes in a uniform grid layout. The grid also presents the range of possible values of each input variable divided into a number of divisions (cells in the grid). Risk Explorer uses colors to convey the risk of unacceptable outcomes (in this case the probability of negative NPVs that will make a loss). The color of each cell in the grid represents the risk of unacceptable outcomes associated with the decision alternative based on the variable's value shown in the cell. Yellow means no risk (i.e. the probability of unacceptable outcomes = 0). Dark orange represents the highest risk (i.e. the probability of unacceptable outcomes = 1). Color was chosen for the purpose of presenting the risk of unacceptable outcomes because it is widely used in risk visualisation as a means to attract people's attention. Also, it is an important visual attention guide that can highlight levels of risk [30]. The numerical values of the risk of unacceptable outcomes can be retrieved by hovering the mouse over the cells (see the pop-up window in Fig. 3).

Risk Explorer also displays the range of possible outcomes resulting from uncertainty in the input variables as a horizontal red/green bar (see Fig. 2(b). The horizontal red/green bar informs the decision-maker about the minimum and maximum potential outcomes under all possible scenarios (i.e. all possible combinations of the variables values). By observing the red part of the bar, the decision-maker can identify the proportion of unacceptable outcomes (in this case, the negative NPVs that will make a loss). Conversely, he/she can identify the proportion of acceptable outcomes (in this case, the positive NPVs that will make a profit) by observing the green part of the bar.

### Overview of uncertainty and risk of unacceptable outcomes

Risk explorer provides an overview of all possible scenarios (i.e. possible combinations of values of input variables) and the risk of unacceptable outcomes associated with the decision alternative under these scenarios. By observing the color variation across the cells, the decision-maker can quickly and easily get an overview of all possible scenarios and identify on which scenarios the decision alternative would be risky. In addition, he or she can readily see the values of the input variables for which the decision alternative is likely to be risky or not. Furthermore, by looking at the displayed range of colors that represents the distribution of risk of unacceptable outcomes, the decision-maker can recognize the trends of the possible risk values (i.e. probabilities of unacceptable outcomes), as well as their relationships with the uncertainty in the input variables. The decision-maker can use this overview to evaluate and compare alternatives in terms of the risk involved in each alternative before focusing on a specific set of scenarios.

In addition to the overview of the risk of unacceptable outcomes, Risk Explorer allows the decision-maker to focus on a specific set of scenarios (i.e. specific values of input variables) and explore the risk of unacceptable outcomes under these scenarios. This can be done by clicking on the cell containing a specific value of one of the input variables, which in turn will open up a new grid showing the new range of risk of unacceptable outcomes with this value fixed. Values of other input variables in the new grid can also be fixed (see Fig. 4).

### Extracting scenarios and keeping history of exploration

Risk Explorer also allows the decision-maker to extract scenarios and maintain a history of the explored scenarios. During the analysis and exploration of scenarios, the decision-maker can select a scenario by specifying the combination of values of input variables that represent the scenario. Then, he or she can click on the "Extract" button to add the selected scenario into the Scenarios Table (see Fig. 4). The Scenarios Table enables the decision-maker to analyse and compare alternatives under a small and more focused set of scenarios.

Extracting and maintain a history of the explored scenarios is particularly useful, as in the presence of many possible scenarios, the decision-maker can be overwhelmed by the amount of information available. Thus, he or she may become unable to remember the scenarios that have been explored, and extract useful information for evaluating and comparing alternatives.

### Analysis and comparison of alternatives at different levels of detail

Risk Explorer enables the decision-maker to focus on a specific set of scenarios (i.e. specific values of input variables) and analyse alternatives under these scenarios. To focus on a specific scenario, the decision-maker needs to fix the values of input variables that represent the scenario. This can be done by clicking on the cell containing a specific value of one of the input variables, which in turn will open up a new grid showing the

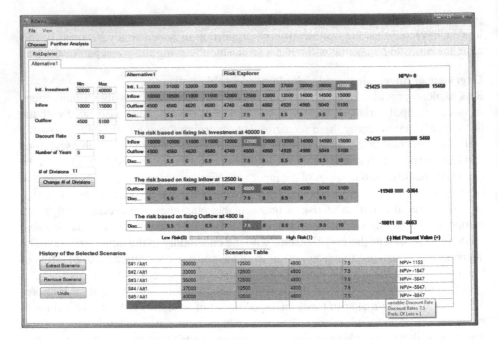

**Fig. 4.** A screenshot of Risk Explorer after analysing Alternative1 under multiple scenarios.

new range of risk of unacceptable outcomes with this value fixed. Values of other input variables in the new grid can also be fixed.

Figure 5 shows an example of analysing alternatives 1 and 2 under specific scenarios based on fixing the two input variables initial investment at $35000 and discount rate at (10%). As shown in Fig. 5, the first fixed value of $35000 in the top grid is highlighted and a new grid is shown for each alternative. The new grid shows the risk values for the other input variables. The risk values are calculated by fixing the values in the high-lighted cells and taking every possible value of the other input variables and calculating what proportion of these combinations will result in unacceptable outcomes. This process is then repeated by fixing the discount rate to 10% in the second grid. In addition to the resulting grid, a new red/green bar is shown to the right of the grid for each alternative. The red/green bar shows the range of possible outcomes resulting from fixing the variables' values in the highlighted cells while varying the other variables within their ranges of values.

Based on the resulting grids and red/green bars, the decision-maker can evaluate and compare alternatives in terms of the risk of unacceptable outcomes and the range of possible outcomes under different scenarios. For example, the new grids and red/green bars in Fig. 5 show that if the two input variables initial investment and discount rate are fixed at $35000 and 10% respectively, then about (78%) of NPVs of alternative 1 (probability 0.777) will result in a loss compared to about 27% (probability 0.273) for alternative 2 (see the popup windows shown in Fig. 5). According to the red/green bars, the maximum loss potential associated with alternative 1 (−$16425) is greater than that associated with alternative 2 (−$8464). In contrast, the maximum potential profit

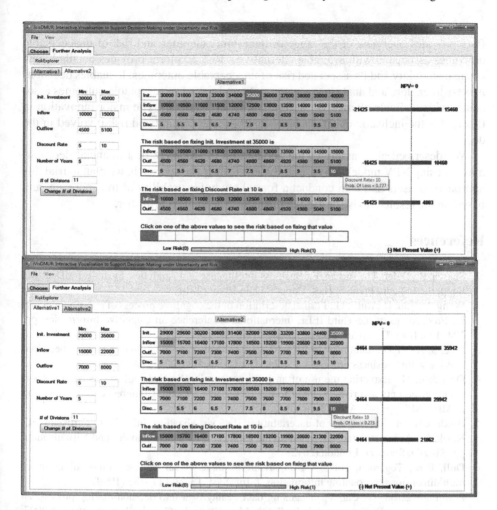

**Fig. 5.** A screenshot of Risk Explorer after exploring alternatives 1 and 2 under initial investment of $35000 and discount rate of 10%.

associated with alternative 1 ($4803) is lower than that associated with alternative 2 ($21862).

# 7 Conclusions and Future Work

In this paper, we presented an information visualisation approach to facilitate the incorporation of uncertainty into the decision-making process called RiDeViz. Our proposed approach to making uncertainty an integral part of decision-making is to view the whole process as one of determining the risk of unacceptable outcomes associated with the decision.

RiDeViz consists of two main interfaces: Decision Bars and Risk Explorer. Decision Bars provides overview of the range of uncertain outcomes and risk of unacceptable outcomes associated with available alternatives. Risk Explorer provides detailed view of the uncertainty and its associated risk of unacceptable outcomes. It enables decision-makers to explore and analyse the uncertainty and risk of unacceptable outcomes associated with available alternatives at different levels of detail. The major innovation in this tool is the inclusion of information about the uncertainty and risk involved in the decision.

We described an example of the application of RiDeViz in a financial decision-making using NPV model as a case study. For testing the approach, we plan to trial this approach on users and then conduct a formal study for usability of the approach, and based on the results, revise or expand the approach and test it further.

# References

1. Sharda, R., Delen, D., Turban, E.: Business Intelligence and Analytics: Systems for Decision Support, 10th edn. Prentice Hall, Upper Saddle River (2014)
2. Wang, Y., Liu, D., Ruhe, G.: Formal description of the cognitive process of decision making. In: Proceedings of the Third IEEE International Conference on Cognitive Informatics. pp. 124–130. IEEE Computer Society (2004)
3. Pereira, A.C.T.D., Santoro, F.M.: A case study on the representation of cognitive decision-making within business process. Int. J. Inf. Decis. Sci. 5(3), 229–244 (2013)
4. Daradkeh, M.: Exploring the use of an information visualization tool for decision support under uncertainty and risk. In: Proceedings of the International Conference on Engineering & MIS 2015, pp. 1–7. ACM, Istanbul (2015)
5. Brodlie, K., et al.: A review of uncertainty in data visualization. In: Dill, J., Earnshaw, R., Kasik, D., Wong, P.C. (eds.) Expanding the Frontiers of Visual Analytics and Visualization, pp. 81–109. Springer, London (2012)
6. Dull, R.B., Tegarden, D.P.: A comparison of three visual representations of complex multidimensional accounting information. J. Inf. Syst. 13(2), 117–131 (1999)
7. Zhu, B., Shankar, G., Cai, Y.: Integrating data quality data into decision-making process: an information visualization approach. In: Smith, M.J., Salvendy, G. (eds.) Human Interface 2007. LNCS, vol. 4557, pp. 366–369. Springer, Heidelberg (2007). doi:10.1007/978-3-540-73345-4_42
8. French, S.: Modelling, making inferences and making decisions: the roles of sensitivity analysis. TOP 11(2), 229–251 (2003)
9. Tufte, E.R.: Visual Explanations: Images and Quantities, Evidence and Narrative. Graphics Press, Cheshire (1997)
10. Pang, A., Wittenbrink, C., Lodha, S.: Approaches to uncertainty visualization. Vis. Comput. 13(8), 370–390 (1997)
11. Davis, T.J., Keller, C.P.: Modelling and visualizing multiple spatial uncertainties. Comput. Geosci. 23(4), 397–408 (1997)
12. Amar, R.A., Stasko, J.T.: Knowledge precepts for design and evaluation of information visualizations. IEEE Trans. Visual Comput. Graphics 11(4), 432–442 (2005)
13. Olston, C., Mackinlay, J.D.: Visualizing data with bounded uncertainty. In: Proceedings of the IEEE Symposium on Information Visualization (InfoVis 2002), p. 37. IEEE Computer Society (2002)

14. Daradkeh, M., McKinnon, A., Churcher, C.: Visualisation tools for exploring the uncertainty-risk relationship in the decision-making process: a preliminary empirical evaluation. In: Proceedings of the Eleventh Australasian Conference on User Interface, vol. 106, pp. 42–51. Australian Computer Society, Inc., Brisbane (2010)

15. Johnson, C., et al.: NIH-NSF Visualization Research Challenges Report. IEEE Press, Los Alamitos (2005)

16. Maier, H.R., et al.: Chapter Five: Uncertainty in environmental decision making: issues, challenges and future directions. In: Jakeman, A.J., Voinov, A.E., Rizzoli, A.E., Chen, S. (eds.) Environmental Modelling and Software and Decision Support – Developments in Integrated Environmental Assessment (DIEA), vol. 3, pp. 69–85. Elsevier, The Netherlands (2008)

17. Hammond, J.S., Keeney, R.L., Raiffa, H.: Smart Choices: A Practical Guide To Making Better Decisions. Harvard Business School Press, Boston (1999)

18. Halpern, J.Y.: Reasoning about Uncertainty. The MIT Press, Cambridge (2003)

19. Dong, X., Hayes, C.C.: Uncertainty visualizations: helping decision makers become more aware of uncertainty and its implications. J. Cogn. Eng. Decis. Making **6**(1), 30–56 (2012)

20. Dayananda, D., et al.: Capital Budgeting: Financial Appraisal of Investment Projects. Cambridge University Press, Cambridge (2002)

21. Jovanovic, P.: Application of sensitivity analysis in investment project evaluation under uncertainty and risk. Int. J. Project Manage. **17**(4), 217–222 (1999)

22. Tziralis, G., et al.: Holistic investment assessment: optimization, risk appraisal and decision making. Manag. Decis. Econ. **30**(6), 393–403 (2009)

23. Magni, C.A.: Investment decisions, net present value and bounded rationality. Quant. Financ. **9**(8), 967–979 (2009)

24. Kohlhammer, J., May, T., Hoffmann, M.: Visual analytics for the strategic decision making process. In: Amicis, R.D., Stojanovic, R., Conti, G. (eds.) GeoSpatial Visual Analytics. NATO Science for Peace and Security Series C: Environmental Security, pp. 299–310. Springer, Dordrecht (2009)

25. Weber, E.U., Johnson, E.J.: Decisions under uncertainty: psychological, economic and neuroeconomic explanations of risk preferences. In: Glimcher, P.W., et al., (eds.) Neuroeconomics – Decision Making and the Brain. Elsevier (2009)

26. French, S.: Decision Theory: An Introduction to the Mathematics of Rationality. Ellis Horwood, Chichester (1986)

27. Luce, R.D., Raiffa, H.: Games and Decisions: Introduction and Critical Survey. Wiley, New York (1957)

28. Clemen, R.T., Reilly, T.: Making Hard Decisions with DecisionTools, 2nd edn. Duxbury Thomson Learning, Pacific Groce (2001)

29. Marco, B., et al.: Simulation optimization: applications in risk management. Int. J. Inf. Technol. Decis. Mak. (IJITDM) **07**(04), 571–587 (2008)

30. Bostrom, A., Anselin, L., Farris, J.: Visualizing seismic risk and uncertainty. Ann. N. Y. Acad. Sci. **1128**, 29–40 (2008). (Strategies for Risk Communication Evolution, Evidence, Experience)

# Process Analytics Through Event Databases: Potentials for Visualizations and Process Mining

Pavlos Delias[✉] and Ioannis Kazanidis

Eastern Macedonia and Thrace Institute of Technology, 65110 Kavala, Greece
pdelias@teiemt.gr

**Abstract.** Events, routinely broadcasted by news media all over the world, are captured and get recorded to event databases in standardized formats. This wealth of information can be aggregated and get visualized with several ways, to result in alluring illustrations. However, existing aggregation techniques tend to consider that events are fragmentary, or that they are part of a strictly sequential chain. Nevertheless, events' occurrences may appear with varying structures (i.e., others than sequence), reflecting elements of a larger, implicit process. In this work, we propose several transformation templates to a enable a process perspective for raw event data. The basic idea is to transform event databases into a format suitable for process mining (aka event log) to enable the rich toolbox of process mining tools. We present our approach through the illustrative example of the events that happened in Greece during the referendum period (summer 2015).

**Keywords:** Event data · Process mining · Event analytics

## 1 Introduction

Event data play an important role in analyzing interactions at a macro-level for decades. Perhaps the first systematic approach is WEIS [29] (although Mc Clelland himself had provided an undeveloped introduction already since the 60s [28]), which allowed a departure from the idiographic study of political events, towards the use of quantitative methods. The importance of having a systematic method to listen and to record political events, was that big, that different initiatives (including projects funded by Defense Advanced Research Projects Agency, and National Science Foundation) produced systems like KEDS [14], CAMEO [15], ICEWS [32], and recently GDELT [24]. The holy grail of all these efforts is the discovery of regularities of events occurrences.

The aim of this work is to enrich the toolbox of analysts that look for this kind of regularities, by enabling a process perspective (and thus the relevant visualizations) for event data. Discovering knowledge through event databases becomes more and more complex, and interesting, since the primary obstacle in acquiring such data (manual collection and coding procedures) is obsolete. Currently, virtually all (political) articles that can be spotted by web crawlers are

© Springer International Publishing AG 2017
I. Linden et al. (Eds.): ICDSST 2017, LNBIP 282, pp. 88–100, 2017.
DOI: 10.1007/978-3-319-57487-5_7

machine-coded, and are getting registered. Therefore, the challenge has shifted to the analysis side.

Lately, many event analytics techniques with appealing visualizations have been developed (see Sect. 2 for a brief overview). However, these (rich) techniques seem to univocally focus on the pure sequentiality of events occurrences. The dominant ways to portray regularities is Frequent Sequence (or Pattern) Mining (applied to sequences of temporally ordered events) [17], and (considering the visualization aspects) flow diagrams [25], like Sankey plots, which in the case of small data (i.e., few events per case and limited event alphabet) are producing alluring illustrations.

Nevertheless, real-world news events often do not follow a linear movement that is implied by a sequence, but they have varying structures, and these may reflect elements of a larger process that has many descriptive components [33]. In this work, the ambition is to enable the observation of the events through a process perspective. We advocate that such an enablement will contribute the following: (i) Expose causalities among activities, (ii) Reveal decision points by introducing gateway semantics (flow splitting and joining points), and (iii) Manifest pathways to distinct outcomes.

The key components of our approach are quire straightforward: Since event data standards qualify similar formats (that include source actor, target actor, event, and timestamp), it is clear that any approach that has a slightest aspiration to make an impact, must exploit this standardized format. Therefore, we propose transforming the original format into various options, all of them enabling event data formats to get loaded into *Process Mining* [1] tools. Process mining techniques will eventually allow process models discovery. These (automatically discovered) process models can be effectively illustrated, and conceivably deliver the contributions we claimed in the previous paragraph. In addition, a plethora of decision support methods (already proposed for original process mining applications) will be able to get leveraged for event analytics.

We should note that in our approach, we make the fundamental assumption that the notion of a process is relevant for the realizations of events, i.e., that there is some kind of rational structure over the events, and therefore the challenge is to unveil this structure. Therefore, our approach follows the prescriptive paradigm that suggests discovering a model, suitable for a given situation in a particular context, and does not intend to be general (like, for instance, in a descriptive approach, which would have aimed at deriving global laws from the observed phenomena).

Relevant references for the process mining applications that can provide decision support are presented in Sect. 2, along with a brief overview of event sequence visualizations techniques, and the GDELT project. In Sect. 4, we describe the background of the story that we are going to use to illustrate our proposal, and we provide some simple event visualizations for the aggregated data. Our proposal about how a process orientation can be infused into raw event datasets is presented in Sect. 3, while a short discussion concludes the paper.

## 2   Literature Review

The "Global Data on Events, Location, and Tone" (GDELT) project, supported by Google, consists of hundreds of millions event records, extracted from broadcast, print, and online news sources from all over the world [24]. GDELT provides a free, rich, comprehensible, daily updated, CAMEO-coded [15] dataset, which applies the TABARI system [7] to each article to extract all the events disclosed therein.

This valuable resource has been exploited in many research efforts (e.g., [22,23,34,37]), however, event data are treated either as input to ordinary analytics techniques, or as time-series data [21]. Nevertheless, the coding format of GDELT allows for virtually any aggregation method of event data (i.e., per actor, per action, per period, or per location) [40] to be applied. With respect to our knowledge, this is the first time that a framework that enables a process perspective on event data is proposed. This is a significant contribution to the field, since a whole novel family of methods will be enabled.

Indeed, treating event data like sequences, can yield effective visualizations that could support decision makers. Several methods for querying, filtering, and clustering multiple event sequences have been proposed, for example [12,36,39], or the works [18,38] that can handle much larger numbers of event sequences and provide effective visualizations for their aggregations. Moreover, when these methods can be combined in a user-friendly dashboard, decision support can be further improved [16]. For a concise description of how event information can be modeled, retrieved, and analyzed, the reader is directed to [20].

Nevertheless, if we assume a process perspective (i.e., that events are not happening in random, but their occurrence is part of a larger, implicit process), the *process mining* paradigm is enabled, and it would substantially augment the decision support potentials. In particular, the following competences will be facilitated:

- Discover complex structures of events (splitting and merging points, long-distance and multi-step causalities, etc.) even when the process is drifting over time (e.g., due to recurring seasonal effects) [27].
- A family of process mining techniques aims at detecting and explaining differences between executions that lead to different outcomes. Under the general term deviance mining [31], we can see for instance, approaches that return in natural language the significant differences between traces that lead to a special outcome and traces that don't [6], or point out the factors that differentiate the flows [9,10]. Similar behavior (which eventually leads to similar results) can be also identified through trace clustering [8,35], where trace profiles are created based on control-flow, organizational, performance or additional criteria, and then traces are grouped according to some similarity metric. Trace clustering techniques are particularly useful to unclutter process models when a lot of variation exists.
- Check deviation from expected pathways. Another type of process mining is conformance checking, where an existing (ideal) process model is compared to

the event log of the same process. Conformance checking can be used to check if reality, as recorded in the log, conforms to the model and vice versa [2]. Local or global diagnostics can be created to check deviations form the expected pathways, or even to check the effect that possible "history modifications" would have to the discovered model [3].
- Putting timestamped events into a process structure allows to observe the temporal evolution (performance) of the process. In process mining, this family of methods is known as performance analysis [4,5], and can respond to questions like: how process performance evolves? Are some events delayed due to bottlenecks? Would the resolution of bottlenecks or following some special paths accelerate some events [30]?

## 3  Enabling Event Data for Process Mining Framework

The standard format of raw event data (e.g., ICEWS, GDELT) comprises at least the following basic elements: A timestamp, two fields indicating the involved actors (the one treated as the source, and the second as the target), and a code for the pertinent action that took place. For example, Table 1 shows a snippet of a GDELT record, that happened on 2nd of July 2015, and involved Greece (Actor code: GRC) and refugees from Syria (Actor Code: SYRREF). The event that took place is the 043: "Host or receive a visitor at residence, office or home country", but it doesn't refer to the actual arrivals of refugees (it would be quite ironic to call that a "visit"), but to actors from the popular TV show "Game of Thrones" that visited Greece to Call on EU Leaders to Help Refugees Stranded in Greece (the original news can be found at: http://people.com/tv/game-of-thrones-stars-call-on-eu-leaders-to-help-refugees-stranded-in-greece/). Many additional fields (e.g., the location of the event, its tone) are recorded as well.

**Table 1.** A sample record of GDELT (only basic fields are showing)

| SQLDATE | Actor1 code | Actor2 code | Event code |
|---------|-------------|-------------|------------|
| 20150702 | GRC | SYRREF | 043 |

However, this format can not be directly matched to a format that will enable a process perspective, such as the input requirements of Process Mining. More specifically, Process Mining requires at least three basic fields for every record: a *case ID* (to correlate events with a particular case), a *timestamp*, and an *activity* [1]. All applications of Process Mining on different types of data, from low-level machine or software logs [19,26] to incidents' status changes in CRM systems [11], assume this data format. With respect to authors' knowledge, this is the first work that enables event data to get exploited by process mining tools, through the following mappings: As long as it concerns the timestamp field, the pairing is clear: it can be directly matched to the corresponding field of the raw event data. However, it is not clear at all what will be the *case ID*, and/or the *activity*. In Table 2, we provide six alternative mappings that can be applied, each of them delivers a different view of the data, yet all of them administer a process-oriented view.

**Table 2.** Alternative transformations to match raw event data to process mining input requirements.

|   | Case ID | Activity |
|---|---|---|
| 1 | SourceActor_TargetActor | Event code |
| 2 | SourceActor | Event code |
| 3 | TargetActor | Event code |
| 4 | Event code | SourceActor |
| 5 | Event code | TargetActor |
| 6 | Event code | SourceActor_TargetActor |

The first alternative transformation uses the combinations of the two actors (hence the underscore as the joining delimiter) to distinguish the case identifiers, and the event code field to create the alphabet of activities. This way, a case is a bilateral relationship of two actors, which leaves a trace of events that happened and involved both of them. By transforming raw event data in this format, we can use the dataset as an input for automated process discovery.

The next two proposed alternative transformations are similar, in the sense that they still make use of the event code as the activity, and an actor as the case ID, yet these transformations use a single actor (i.e., not a combination). These alternatives are suitable for datasets where many countries participate, and there is a limited thematic selection. In such situations, we expect process maps to reveal if there are any behavioral patterns for single countries (e.g., if countries that "criticize or denounce" are "using conventional military force" as a following action).

Transformations 4, 5, and 6 inverse the logic. Traces are now joined by the event code (case ID). The items that each trace comprise (activity) are either single actors (source or target) or their combinations. This inverse logic allows to observe the interactions of actors subject to a special thematic. In Sect. 5 we provide illustrative instantiations for the relevant templates.

## 4 An Illustrative Example for Event Analytics

July of 2015 was no ordinary month for Greece. Following a dubious negotiation strategy, the Prime Minister Alexis Tsipras, on 27th of June 2015, announced a referendum to decide whether Greece was to accept the bailout conditions, proposed by the European Commission, the European Central Bank, and the

International Monetary Fund. Although the result was a clear "No" (61.3%), one week later, Greece's parliament signed on to harsher bailout terms, i.e., a new mid-term Memorandum of Understanding. News was coming rapid and spectacular, since the government soon called an early parliamentary election. The story has been extensively covered by international media, making it eminently suitable for an illustrative example for the proposed approach.

**Fig. 1.** Geographic locations of the recorded articles (heatmap).

Dozens of thousands of events were registered to the GDELT, allowing for rich and informative event analytics. Several techniques and visualizations are provided through the GDELT Analysis Service (http://analysis.gdeltproject.org). In Figs. 1, 2, and 3 we present some indicative visualizations of the GDELT A.S. that illustrate parts of the story. In particular, Fig. 1 presents a geographic heatmap to understand the spatial patterns of the events that were realized between 20th June 2015–20th July 2015, having as the actor's country Greece. The heatmap weighs each location by the total number of unique events found at that location, irrespective of how much news coverage each event received. It comes at no surprise that besides Athens, most events of that period (initiated by Greece) happened at the great capital cities of the European Union (Brussels, Berlin, Paris, London, etc.), i.e., the "decision centers" of E.U. The heatmap has also spotlighted Lesvos island, due to the refugees movements that were at full deployment during that period.

Moreover, GDELT constructs the so-called Global Knowledge Graph (GKG) by extracting information such as the connections of persons, organizations, locations, emotions, and themes identified in the source texts. A word cloud of the most popular themes identified in the events that contained the keyword "Greece" and happened between 20th June 2015–20th July 2015 is illustrated in Fig. 2. Unsurprisingly, we see the term "tax" to dominate the picture. Other terms that outstand are "debt", "bankruptcy", "negotiation", "fragility", which accurately reflect the situation of that period. The way that GDELT A.S. generate the input data, the word cloud essentially weights each object towards those that occur in the greatest diversity of contexts (to make it insusceptible to sudden massive bursts of coverage).

**Fig. 2.** Word cloud for the themes of the articles of the referenced period.

The emotions registered in the GKC can also provide what is called the "tone" (a score ranges from −100 (extremely negative) to +100 (extremely positive) that denotes negative/positive emotional connotation of the words in the article). In Fig. 3, we present a timeline of the smoothed daily average tone for all events that contained the keyword "Greece" and happened between 20th June 2015–20th July 2015. We have highlighted with red color the dates of the announcements for the referendum and the early elections, at which point we regard two sudden drops of the tone (the negative emotions become even more negative).

All these interesting visualizations provide interesting insights, derived from various aggregation perspectives of the events. However, in all of them, events are considered fragmentary, "rambling" elements of the story. Should anyone develop a hypothesis that events occur as part of a regularity, i.e., a process, these kind of visualizations can not support the relevant checking. In the next section, we present our proposal for a general framework about enabling a process perspective for the same event datasets. As the sample dataset, we will use

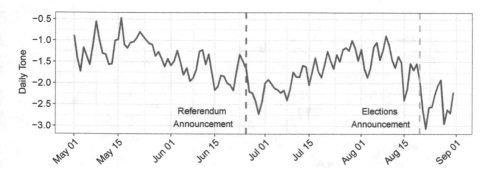

**Fig. 3.** Tone timeline for the referenced period. (Color figure online)

all events that happened between 20th June 2015–20th July 2015, and whose actors' country (either source actor's or target actor's) is Greece. Moreover, we considered the events that happened in Belgium (host country of the EU institutions, where most official negotiations took place) during the same period. We should stress that this dataset is used for illustrative purposes, and that we do not claim it to be a complete set.

## 5  An Application Beyond Event Analytics: Getting Insights from Event Data Following a Process Perspective

In this section, we apply the transformation templates presented in Sect. 3 and we depict the discovered process maps for the indicative dataset described in Sect. 4 (using Disco ® [13]) in Figs. 4a to d. In general, darker boxes indicate more frequent activities, and directed arrows indicate the transitions among them. Again, infrequent transitions have been filtered out. Although Fig. 4 is used for illustrative purposes, and we do not claim a full and complete analysis of the case study story, several patterns become evident. For instance, as we can see in Fig. 4a (which illustrates the first transformational template), a common pattern in bilateral relationships is to "host a visit", then either to "make a pessimistic comment" or to "express intent to meet or negotiate", in which case what follows is an "engagement in diplomatic cooperation", a pattern that competently reflects the turbulent political situation of that time.

The transformations 2 and 3 (SourceActor or TargetActor as Case IDs, and Event Code as the Activity) are suitable for datasets where many countries participate, and there is a limited thematic selection (not the case we described in Sect. 4), so we don't actually provide any visualizations for them.

To capture the interactions of actors subject to a special thematic, we may use the transformations 4, 5, and 6. Of particular interest for our illustrative example is the sixth transformation. We demonstrate this kind of potentials with Fig. 4b, c, and d. Figure 4b presents what pairs of actors were involved in "Make

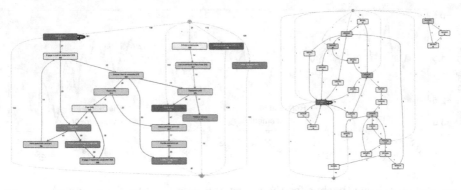

(a) Process map for the first transformation alternative

(b) Process map for the "Public Statement" thematic -6th transformation

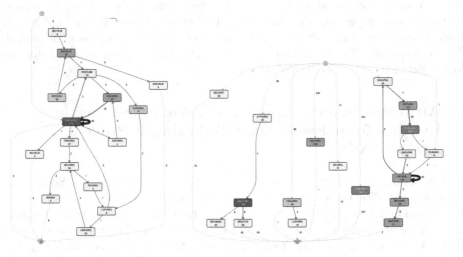

(c) Process map for the"Disapprove"thematic-6thtransformation

(d) Process map for the "Consult" thematic -6th transformation

**Fig. 4.** Process maps

Public Statement" events (e.g., make pessimistic/optimistic comment, acknowledge or claim/deny responsibility, make empathetic comment). Two interesting patterns that emerge are (*i*) the loop over the GRC//GRC box, which commonly signifies a ping-pong of comments between the government and the opposition, and (*ii*) the path that connects the pair of Greece and EU, followed by Greece and USA, followed by Greece and Germany, followed by the pair Germany and Greece (the order is important since we assume different roles for actors: source vs. target). It is a path that reflects a common situation of that time:

Greek authorities make a statement about the opinions/decisions of EU institutions, then make a comment about the IMF, and then watch the protagonists (Greece and Germany) making refined commentaries. Similar patterns are visible for the "Disapprove" thematic (Fig. 4c), which refers to events like criticize or denounce, accuse, rally opposition, complain officially, etc.

A different part of the same story is illustrated in Fig. 4d, which shows the events of the "consult" thematic (i.e., events like discuss by telephone, make/host a visit, meet at "third" location, engage in mediation/negotiation). It is remarkable to see well separated threads: the left-most represents the interactions among the balkan countries Greece, Turkey, and Former Yugoslavic Republic of Macedonia, in the middle we observe one thread for the "consulting" sessions between Greece, France and Luxembourg (the country of Jean-Claude Juncker, President of the European Commission, who played a key-role in the negotiations), as well as one distinguished thread for Germany and Greece (the protagonists). At the right, there is also a path that involves multiple actors. Alternative transformations may be relevant for some situations (e.g., joining event code and actor either for the case id field or for the activity field), however, the goal of this paper was to exhibit the advantages of enabling a process perspective for the raw event data. Hopefully, the power of representing events on a process map is evident in all the cases we discussed.

## 6 Discussion

In this work, we tried to exploit the richness and availability of raw event data by adding analytics capabilities. These capabilities are allowed by the enablement of a process perspective for the (otherwise disjointed) events. Although we didn't discuss explicitly all the advantages that a process representation can yield, we hope that the general framework, and the process maps we presented are promising and indicate the expected benefits. In particular, the illustrative example emphasized the visualization of diverging and converging of the flow into distinct paths through the splitting and merging points, as well as the discovery of multi-step causalities. After the data transformation, it would be possible to apply additional process analytics, like those described in the second part of Sect. 2, however, mainly due to space limitations, such analytics are left out of this work.

There are of course, several limitations in our approach. A critical one is that, by definition, a process has definitive start and end points. However, in every application one needs to justify very well and very carefully the assumption that international relationships can fit to this requirement. It is within our future plans to test practices from data stream analysis to support dealing with this limitation. Anyway, this paper puts forward a big promise for event analytics, and many challenges may appear, however, given the efforts that have already been devoted to data collection issues, the focus needs to be shifted towards the analysis side.

# References

1. van der Aalst, W.: Process Mining: Data Science in Action, 2nd edn. Springer, Heidelberg (2016). doi:10.1007/978-3-662-49851-4
2. Van der Aalst, W., Adriansyah, A., van Dongen, B.: Replaying history on process models for conformance checking and performance analysis. Wiley Interdisc. Rev. Data Min. Knowl. Discovery **2**(2), 182–192 (2012)
3. van der Aalst, W.M., Low, W.Z., Wynn, M.T., ter Hofstede, A.H.: Change your history: learning from event logs to improve processes. In: 2015 IEEE 19th International Conference on Computer Supported Cooperative Work in Design (CSCWD), pp. 7–12. IEEE (2015)
4. Van der Aalst, W.M., Schonenberg, M.H., Song, M.: Time prediction based on process mining. Inf. Syst. **36**(2), 450–475 (2011)
5. Adriansyah, A., Buijs, J.C.A.M.: Mining process performance from event logs. In: Rosa, M., Soffer, P. (eds.) BPM 2012. LNBIP, vol. 132, pp. 217–218. Springer, Heidelberg (2013). doi:10.1007/978-3-642-36285-9_23
6. Beest, N.R.T.P., Dumas, M., García-Bañuelos, L., Rosa, M.: Log delta analysis: interpretable differencing of business process event logs. In: Motahari-Nezhad, H.R., Recker, J., Weidlich, M. (eds.) BPM 2015. LNCS, vol. 9253, pp. 386–405. Springer, Cham (2015). doi:10.1007/978-3-319-23063-4_26
7. Best, R.H., Carpino, C., Crescenzi, M.J.: An analysis of the tabari coding system. Confl. Manag. Peace Sci. **30**(4), 335–348 (2013)
8. Bose, R.J.C., van der Aalst, W.M.: Context aware trace clustering: towards improving process mining results. In: SDM, pp. 401–412. SIAM (2009)
9. Leoni, M., Aalst, W.M.P., Dees, M.: A general framework for correlating business process characteristics. In: Sadiq, S., Soffer, P., Völzer, H. (eds.) BPM 2014. LNCS, vol. 8659, pp. 250–266. Springer, Cham (2014). doi:10.1007/978-3-319-10172-9_16
10. Delias, P., Grigori, D., Mouhoub, M.L., Tsoukias, A.: Discovering characteristics that affect process control flow. In: Linden, I., Liu, S., Dargam, F., Hernández, J.E. (eds.) Decision Support Systems IV-Information and Knowledge Management in Decision Processes. Lecture Notes in Business Information Processing, vol. 221, pp. 51–63. Springer, Cham (2015)
11. van Dongen, B., Weber, B., Ferreira, D., De Weerdt, J.: Proceedings of the 3rd Business Process Intelligence Challenge (Co-located with 9th International Business Process Intelligence Workshop, BPPI 2013, Beijing, China, 26 August 2013
12. Fails, J.A., Karlson, A., Shahamat, L., Shneiderman, B.: A visual interface for multivariate temporal data: finding patterns of events across multiple histories. In: 2006 IEEE Symposium on Visual Analytics Science and Technology, pp. 167–174. IEEE (2006)
13. Fluxicon: Disco. Fluxicon (2012). http://www.fluxicon.com/disco/
14. Gerner, D.J., Schrodt, P.A., Francisco, R.A., Weddle, J.L.: Machine coding of event data using regional and international sources. Int. Stud. Q. **38**(1), 91–119 (1994)
15. Gerner, D.J., Schrodt, P.A., Yilmaz, O., Abu-Jabr, R.: Conflict and Mediation Event Observations (Cameo): A New Event Data Framework for the Analysis of Foreign Policy Interactions. International Studies Association, New Orleans (2002)
16. Gotz, D., Stavropoulos, H.: DecisionFlow: visual analytics for high-dimensional temporal event sequence data. IEEE Trans. Vis. Comput. Graph. **20**(12), 1783–1792 (2014). doi:10.1109/tvcg.2014.2346682. http://dx.doi.org/10.1109/TVCG.2014.2346682

17. Gotz, D., Wang, F., Perer, A.: A methodology for interactive mining and visual analysis of clinical event patterns using electronic health record data. J. Biomed. Inf. **48**, 148–159 (2014). doi:10.1016/j.jbi.2014.01.007. http://dx.doi.org/10.1016/j.jbi.2014.01.007

18. Gotz, D., Wongsuphasawat, K.: Interactive intervention analysis. In: AMIA Annual Symposium Proceedings, vol. 2012, pp. 274–280. American Medical Informatics Association, Washington, DC, USA (2012)

19. Günther, C.W., Rozinat, A., Aalst, W.M.P.: Activity mining by global trace segmentation. In: Rinderle-Ma, S., Sadiq, S., Leymann, F. (eds.) BPM 2009. LNBIP, vol. 43, pp. 128–139. Springer, Heidelberg (2010). doi:10.1007/978-3-642-12186-9_13

20. Gupta, A., Jain, R.: Managing event information: modeling, retrieval, and applications. Synth. Lect. Data Manag. **3**(4), 1–141 (2011)

21. Jiang, L., Mai, F.: Discovering bilateral and multilateral causal events in GDELT. In: International Conference on Social Computing, Behavioral-Cultural Modeling & Prediction (2014)

22. Keertipati, S., Savarimuthu, B.T.R., Purvis, M., Purvis, M.: Multi-level analysis of peace and conflict data in GDELT. In: Proceedings of the MLSDA 2014 2nd Workshop on Machine Learning for Sensory Data Analysis, p. 33. ACM (2014)

23. Kwak, H., An, J.: Two tales of the world: comparison of widely used world news datasets GDELT and eventregistry (2016). arXiv preprint arXiv:1603.01979

24. Leetaru, K., Schrodt, P.A.: GDELT: global data on events, location and tone, 1979–2012. In: Resreport International Studies Association, Graduate School of Library and Information Science, University of Illinois at Urbana-Champaign, USA (2013). http://data.gdeltproject.org/documentation/ISA.2013.GDELT.pdf

25. Liu, Z., Wang, Y., Dontcheva, M., Hoffman, M., Walker, S., Wilson, A.: Patterns and sequences: interactive exploration of clickstreams to understand common visitor paths. IEEE Trans. Vis. Comput. Graph. **23**(01), 321–330 (2017)

26. Mannhardt, F., Leoni, M., Reijers, H.A., Aalst, W.M.P., Toussaint, P.J.: From low-level events to activities - a pattern-based approach. In: La Rosa, M., Loos, P., Pastor, O. (eds.) BPM 2016. LNCS, vol. 9850, pp. 125–141. Springer, Cham (2016). doi:10.1007/978-3-319-45348-4_8

27. Martjushev, J., Bose, R.P.J.C., Aalst, W.M.P.: Change point detection and dealing with gradual and multi-order dynamics in process mining. In: Matulevičius, R., Dumas, M. (eds.) BIR 2015. LNBIP, vol. 229, pp. 161–178. Springer, Cham (2015). doi:10.1007/978-3-319-21915-8_11

28. McClelland, C.A.: The acute international crisis. World Polit. **14**(01), 182–204 (1961)

29. McClelland, C.A.: World event/interaction survey codebook (1976)

30. Nguyen, H., Dumas, M., Hofstede, A.H.M., Rosa, M., Maggi, F.M.: Business process performance mining with staged process flows. In: Nurcan, S., Soffer, P., Bajec, M., Eder, J. (eds.) CAiSE 2016. LNCS, vol. 9694, pp. 167–185. Springer, Cham (2016). doi:10.1007/978-3-319-39696-5_11

31. Nguyen, H., Dumas, M., Rosa, M., Maggi, F.M., Suriadi, S.: Mining business process deviance: a quest for accuracy. In: Meersman, R., Panetto, H., Dillon, T., Missikoff, M., Liu, L., Pastor, O., Cuzzocrea, A., Sellis, T. (eds.) OTM 2014. LNCS, vol. 8841, pp. 436–445. Springer, Heidelberg (2014). doi:10.1007/978-3-662-45563-0_25

32. O'Brien, S.P.: Crisis early warning and decision support: contemporary approaches and thoughts on future research. Int. Stud. Rev. **12**(1), 87–104 (2010)

33. Peuquet, D.J., Robinson, A.C., Stehle, S., Hardisty, F.A., Luo, W.: A method for discovery and analysis of temporal patterns in complex event data. Int. J. Geograph. Inf. Sci. **29**(9), 1588–1611 (2015). doi:10.1080/13658816.2015.1042380. http://dx.doi.org/10.1080/13658816.2015.1042380
34. Phua, C., Feng, Y., Ji, J., Soh, T.: Visual and predictive analytics on Singapore news: experiments on GDELT, Wikipedia, and ˆsti (2014). http://arxiv.org/abs/1404.1996
35. Song, M., Günther, C.W., Aalst, W.M.P.: Trace clustering in process mining. In: Ardagna, D., Mecella, M., Yang, J. (eds.) BPM 2008. LNBIP, vol. 17, pp. 109–120. Springer, Heidelberg (2009). doi:10.1007/978-3-642-00328-8_11
36. Vrotsou, K., Johansson, J., Cooper, M.: Activitree: interactive visual exploration of sequences in event-based data using graph similarity. IEEE Trans. Vis. Comput. Graph. **15**(6), 945–952 (2009)
37. Ward, M.D., Beger, A., Cutler, J., Dickenson, M., Dorff, C., Radford, B.: Comparing GDELT and ICEWS event data. Analysis **21**, 267–297 (2013)
38. Wongsuphasawat, K., Gotz, D.: Exploring flow, factors, and outcomes of temporal event sequences with the outflow visualization. IEEE Trans. Vis. Comput. Graph. **18**(12), 2659–2668 (2012). doi:10.1109/tvcg.2012.225. http://dx.doi.org/10.1109/TVCG.2012.225
39. Wongsuphasawat, K., Plaisant, C., Taieb-Maimon, M., Shneiderman, B.: Querying event sequences by exact match or similarity search: Design and empirical evaluation. Interact. Comput. **24**(2), 55–68 (2012)
40. Yonamine, J.E.: Working with event data: a guide to aggregation choices. Penn State University: Working Paper (2011)

# Value of Visual Analytics to South African Businesses

Wisaal Behardien and Mike Hart[(✉)]

Department of Information Systems, University of Cape Town, Cape Town, South Africa
wisaalbehardien@gmail.com, mike.hart@uct.ac.za

**Abstract.** There is limited literature on the value that visual analytics provides for businesses, and its broad use in organisations. This research provides some understanding of how South African businesses are using visual analytics in their day to day operations, and the value derived from employing it. The study was interpretive, exploratory and descriptive, producing both quantitative and qualitative data. Individuals within organisations making use of visual analytics completed an online survey, and interviews were conducted with informed business, IT and BI stakeholders. Results were compared with those from an international survey, and thematic analysis highlighted four main themes: usage, value, challenges and technology. Most respondents noted the high added value obtained from visual analytics versus tables of numbers. The research also identified a set of good practices for organisations to employ when embarking on a visual analytics strategy and suggested ways of mitigating potential challenges.

**Keywords:** Visual analytics · Visualisation · Value · Benefits · Challenges · Business intelligence · Data · Change management

## 1   Introduction

For many years Business Intelligence and Analytics (BI&A) has been rated the largest and most significant IT investment internationally [1, 2] and the top technology priority [3]. However, the percentage of employees in organisations using BI&A is still very limited [4, 5, p. 88]. An essential component of BI&A is the appropriate presentation of data [6]; visual display of data allows users to gain more insight and draw conclusions timeously and easily [7–9]. Businesses find value in data visualisation because many users are more effective at recognising patterns in graphic representations of data than they are with rows of data in tabular format [10].

Visual analytics (VA) encourages effective and smarter decisions within a business [7, 9, 11]. VA incorporates the science of analytical reasoning and involves human cognition [12]. For the purpose of this research visual analytics will refer to the use of visualisation, human factors and data analysis to facilitate interactive visual interfaces of data from BI&A systems [12, 13].

VA research shows various business benefits, but users are not using it to its full potential to improve their decision making and analytical processes [9, 12]. Most VA studies are intended for designers and analysts, and few concern benefits of VA systems for users and managers [14]. The objective of this research is therefore to examine the

© Springer International Publishing AG 2017
I. Linden et al. (Eds.): ICDSST 2017, LNBIP 282, pp. 101–116, 2017.
DOI: 10.1007/978-3-319-57487-5_8

usage of visual analytics in South African business organisations, and obtain some understanding of its value, related challenges, and suggestions for good practice.

Structure of this paper: Some background is given on the role of visual analytics in business, the value offered, and potential challenges. The research methodology then follows. After this, summary results of a questionnaire survey are discussed, together with themes emerging from interviews with practitioners of visual analytics. This is followed by recommendations of good practices for business organisations to follow in using visual analytics, and the paper then concludes.

## 2     Background to the Research

### 2.1     Analytics

Analytics is a subset of business intelligence used to support almost all business processes [15] and a broad term used to cover any data-driven decision [16]. It has three broad categories: descriptive analytics, predictive analytics and prescriptive analytics. From their surveys [17] confirm the great business potential for analytics, but note that it is not easy to obtain sustainable competitive advantage. For decision support purposes, data processed into information needs to be presented in a way most appropriate for the end user [18]. Different technologies can be used to present the data, such as OLAP, dashboards and scorecards, and visual analytics.

### 2.2     Visual Analytics

Over the years academics have used different terms for the phenomenon of representing data graphically, including visualisation, visual analytics, data visualisation, visual data mining, information visualisation, business visualisation, visual analysis and business information visualisation; many are used interchangeably [19]. For the purpose of this study the term visual analytics (VA) will be used, with the frequently–used definition: "Visual analytics is the science of analytical reasoning facilitated by interactive visual interfaces" [13, p. 4].

VA can be divided into three components, namely visualisation, data management and data analysis [20]. Data management integrates different data sources in order to analyse a consolidated view of the data [20]. Data analysis uses algorithms in order to transform the data [12]. Visualisation reduces the complexity of the data by breaking it down to a graphical form [9, 11, 21], using different types of images to represent information to users in a way that assists them to gain insight and draw conclusions [9]. Using VA for storytelling is recommended [22], while [23] describe VA as a translational cognitive science.

Discussing the application of interactive VA to decision-making and problem-solving, [24, p. 20] comment that it "differs from other forms of digital creativity, as it utilizes analytic models, relies on the analyst's mental imagery and involves an iterative process of generation and evaluation of ideas in digital media, as well as planning, execution, and refinement of the associated actions". According to [25], VA is useful to reduce the cognitive work needed by a user to perform a specific task.

## 2.3    Value of Visual Analytics in a Business Context

There is a great need in business to create tools that enable people to synthesise information and derive insight from massive amounts of data, provide understandable assessments of the current state of business, detect unexpected patterns within data and be able to draw conclusions that are a fair representation of the truth [12]. According to [9], business executives value the visualisation of data, and graphical data representations have the potential to be used throughout business organisations.

Well-executed VA can be of great value to businesses as it can turn the information overload into an opportunity, support smarter decisions, and provide a competitive advantage [9, 25]. The fact that VA reduces the cognitive work needed by users to perform tasks means that users have more time to focus on other aspects within the business [25]. "Visual analytics aims to put the human in the loop at just the right point to discover key insights, develop deep understanding, make decisions, and take effective action" [26, p. 328].

## 2.4    Challenges in Visual Analytics

Visualisation challenges are identified by [27] in three areas: human-centered, technical and financial. The first area includes interdisciplinary collaboration, evaluation of usability [28, 29], finding effective visual metaphors, choosing optimal levels of abstraction, collaborative visualisation, effective interaction, and data quality [25, 30]. Technical challenges include scalability [12], high data dimensionality, time-dependent data, data filtering, and platform independent visualisation, and parallelism [30]. It is difficult to analyse the data under multiple perspectives and assumptions to understand both the historical and current situations, and larger data sets are more difficult to manage, analyse and visualise effectively [25].

Limited research has been published on the usage of VA and its broad implementation in business organisations. The objective of this research is to fill this gap in literature by providing some insight into the application of VA in South African businesses. Given space restrictions, the focus will be on organisational and people-related issues rather than technology-related ones.

# 3    Research Methodology

The research adopted an interpretive philosophy and inductive approach, as the main aim was to gain more insight into how South African businesses were using and gaining value from VA. It was cross-sectional and both descriptive and exploratory, using mixed methods. Permission for the research was obtained from the university's research ethics committee, and organisational respondents all gave their agreement.

## 3.1   Questionnaire Survey

Non-probability purposive sampling [31] was used, by contacting organisations experienced in BI&A, and asking if staff involved in producing, deploying or using VA would be prepared to respond to a survey on VA. A questionnaire, based on questions in [9], was drawn up in Qualtrics survey software, with questions modified where necessary to suit the study and the South African context, and additional questions included. A cover letter and link was then emailed to the sample. Responses were checked in Qualtrics for completion, and exported to Microsoft Excel for further calculation and comparison with results from [9]. The intention was not for statistical comparison or testing, but rather to gain a sense of how similar the South African situation was to the international results of [9], and aid overall interpretation. There was some bias towards retailers, as many have their head offices in Cape Town, from where the survey was conducted. Table 1 gives details of the 35 respondents completing the questionnaire.

**Table 1.**   Summary details of respondents to on-line survey

| Industry | No. | Role | No. |
|---|---|---|---|
| Retail/wholesale/distribution | 14 | Data and IT professional | 23 |
| Consulting/professional services | 5 | Business exec/sponsor/user | 6 |
| Financial services | 4 | Consultant | 6 |
| Education | 3 | Total | 35 |
| Healthcare | 2 | | |
| Telecommunications | 2 | **Number of Org. Employees** | **No.** |
| Software/internet services | 1 | More than 100,000 | 2 |
| Government | 1 | 10,000 to 100,000 | 10 |
| Media/entertainment/ publishing | 1 | 1,000 to 9,999 | 11 |
| Chemicals | 1 | 100 to 999 | 6 |
| Oil & gas | 1 | Less than 100 | 6 |
| Total | 35 | Total | 35 |

## 3.2   Semi-structured Interviews

The next step was to conduct thirteen semi-structured interviews in order to gain a deeper understanding of the value that business organisations were deriving from using VA. The interview protocol was designed keeping the Technology-Organisation-Environment Model (TOE) in mind as well as the aspects that the researchers had found in literature [32, 33]. Interviewees came from business organisations making use of VA in some form, and were spread across the organisational spectrum from the technical, analysis and deployment side to users at different levels, as well as a consultant in this area. Interviews were almost all face-to-face and voice recorded, but video conferencing and telephonic interviews had to be conducted in a few cases. Table 2 shows the Job Titles of the thirteen interviewees, with the codes given to them.

**Table 2.** Job titles of interviewees with codes given

| Code | Job title | Code | Job title |
|------|-----------|------|-----------|
| B1 | CEO | H1 | BI developer |
| B2 | Head of market research | H2 | BI analyst |
| B3 | Business analyst | H3 | Head of BI |
| B4 | Business analyst | H4 | BI analyst |
| B5 | Data analyst | H5 | Head of BI |
| B6 | Consultant | H6 | Solution architect |
| B7 | Head of process development | | |

Interviews were transcribed and annotated from the recordings, and supplemented with the researcher's notes. The qualitative data was then analysed by employing the six step procedure for thematic analysis of [34] to describe both the implicit and explicit ideas within the data. Thematic analysis was done by looking for key words, themes or ideas within the data, with codes given to specific themes and sub-themes within the data to capture the complexities of meaning [35].

# 4    Research Findings

This describes the main findings from both the interviews and the questionnaire survey, which have been integrated to obtain as full a picture as possible. The main themes and subthemes emerging from the thematic analysis are shown in Table 3. Note that due to space limitations the theme of technology will not be discussed, and the focus will be more on the usage, value and challenges of VA to the businesses.

**Table 3.** Themes and sub-themes that emerged from thematic analysis of interviews

| Theme | Sub-themes |
|-------|-----------|
| Usage | Business usage, interaction and self-service, visual data discovery, and stakeholder involvement |
| Value | Integration and knowledge sharing, exception highlighting, decision support and faster actionable insight, and saves time |
| Challenges | User acceptance and adoption, lack of effective change management, lack of management support, lack of skills, and uncertainty of data quality |
| Technology | Data, data storage; and visual analytic tools |

## 4.1   Usage of Visual Analytics

**Business Usage.** The percentages of online survey respondents replying that various business areas were using graphical representations of data are shown in Fig. 1.

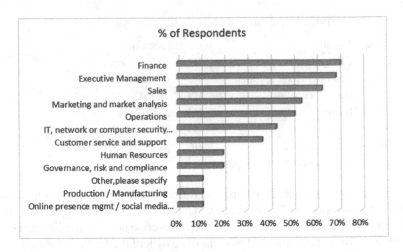

**Fig. 1.** Business areas using graphical representations of data (n = 35)

Interviewees made various points in this regard. B5 stated that their aim was to get *"every level of the company looking at these dashboards."*

H3 commented that in their organisation *"reporting we have over 2000 users so everyone from the store users till the top is exposed to it."*

H4 said it was important for them to have *"role based reporting to ensure that the information being displayed was relevant to the user."*

Many interview participants stated that their marketing departments found VA extremely appealing when analysing customer data. The data suggests that VA is a good way to "convert" business users who are less inclined to analyse numeric data.

Figure 2 compares the percentage of survey respondents in this study with those in [9] who said that VA was being used by various roles. It suggests a relatively good spread of VA across stakeholders and employees in the South African businesses.

*Some points on the data itself:* To the survey question: "what kind of data is being analysed and visually represented within your company" the percentage of survey respondents replying was: structured (91%), semi-structured (31%), unstructured (23%), machine generated (e.g. sensor, RFID) (11%), and social media (3%).

> *"our CEO and sales manager they require an entirely different sort of data view into this. I think for the company at large, for lower levels we are painting broad brush strokes, unless its specific data to their functions. So if its details, team needs to see their targets they need that info. Definitely from CEO and management perspective they need higher level decision making information"* (B5).

**Interaction and Self-service.** These were major reasons for the increased usage of analysis visually. B3 commented that *"I can drilldown, I can start at the highest level and drilldown, and it's cool as I have never had this before. We get excited about pictures and colours."*

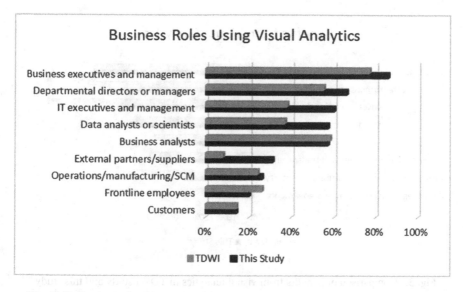

**Fig. 2.** Roles in the business using visual analytics (from two questionnaire surveys)

H2 said: *"Our VA platform supports decisions by allowing users to filter data, drill down etc. so, interact with data"*.

**Visual Data Discovery.** Interviewees highlighted that their organisations ventured into VA in order to create an empowering environment where business users would seek answers to their questions. The researchers observed a desire for VA to promote innovation as well as knowledge in the organisation. H5 commented that they wanted to *"empower users to analyse."* and *"We want innovation, we want users to see what we have given them and come up with innovative solutions to problems that arise"*.

**Stakeholder Involvement.** Stakeholder involvement was a key aspect that interviewees believed necessary for a successful VA strategy. H3 stated that they worked with the business units to ensure that they had all the necessary information before new visualisations were created for that particular business unit.

> *"It is important for us to work closely with business units so that we can ensure that what they are getting is relevant to them so that we don't waste our time or their time"*. (H2)

### 4.2   Value and Benefits of Visual Analytics

Online survey respondents were asked the same question as those in the TDWI survey (Stodder 2013). Figure 3 shows that the percentages of South Africans benefitting in various ways were not all that different from those of the international respondents. The top three items were the same, but South Africa was well behind in improving operational efficiencies, and in increasing employee and partner productivity.

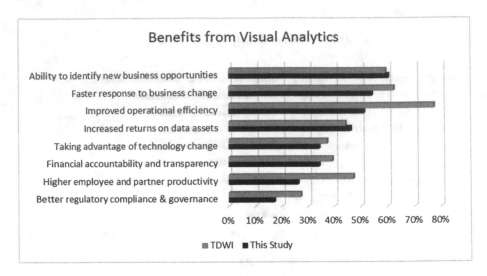

**Fig. 3.** Comparison of benefits from visual analytics in TDWI study and this study

Respondents to the questionnaire surveys were given the specific list of benefits in Fig. 3, whereas interviewees were able in conversation to define the actual value they gained from VA. This produced the following four main sub-themes of Table 3.

**Integration and Knowledge Sharing.** Interviewees felt that use of VA would allow each business unit to integrate better with each other. These often work in separate "silos" from each other, unaware of what is happening in the organisation as a whole. H4 commented that in their organisation *"each business unit has its own silos."* B5 noted that VA would *"promote cross disciplinary use, which would be good at fostering relationships there and getting a better context of the business and getting a better viewing to it"*.

**Exception Highlighting.** Interviewees pointed out that if there is unusual data within a data set, visualisations allow you to identify these exceptions or trends within the data. H4 stated that they *"want to highlight exceptions."*

> *"If you put the right visualisation, I think you can pick up anomalies more easily. Allows you to compare different periods of time,"* (H3).

**Decision Support and Faster Actionable Insight.** This sub-theme drew many comments from interviewees. Participants wanted to have actionable insight; VA provides a platform that allows business users to absorb information quicker and respond faster. VA facilitates root cause analysis through the ability to interact with the data that is displayed, to drill down into a transactional level. H1 commented that VA gave *"faster insight and a good understanding of the business at a high-level."* H3 highlighted that *"it's more about speed of access to information, ..... it's about asking the questions and how quickly and easily I can get to the answer."*

*"So what I think it means for the organisation is more efficiency, more effectiveness in reaching our goals reaching our targets and I think a mind-set shift from anecdotal, information relying on tacit knowledge that's inside the mind of our team members, to data driven decision making which hopefully leads to more effective efficient and better decisions"* (B2).

H5 stated that they *"want innovation, want users to ask questions about the data"* and want to use VA as a way to do *"visual storytelling, going to tell a story with data, what is the problem, use the data to tell the story and use visualisations to come to the conclusion."*

H2 commented that *"people are given more insight into the data"* and that VA *"supports decisions by allowing users to filter data, drill down etc. interact with data."*

In the online survey the following question was posed: *"To what degree do visualisations of data improve business insight?"* Replies were split as follows: Very High (11%), High (51%), Moderate (37%), Low (0%) and Very Low (0%).

Because of concern that the benefits described might be largely due to analytics generally, as opposed to the visual aspect specifically, the following question was also asked in the local survey: *"To what degree do users within the company find value in using visualisation, compared to just numbers and tables?"* Answers were: Very High (17%), High (43%), Moderate (31%), Low (6%) and Very Low (3%), indicating that over 60% felt that visualisation was an improvement over conventional numeric presentation formats.

**Saving Time.** The data suggests that VA saves business users time. Employing VA, according to interviewees, allows organisations to streamline different processes and reduce the total time taken. B5 states *"up until now a lot of the work is automating manual processes, so people have been populating Excel spreadsheets and manipulating data. To cut down the time that they spend on this."*

*"They previously used to take 21 days to get everything ready, and now they have it done within a few hours, so it has saved them so much time and trouble. It also saved them a lot of money as well"* (B3).

*"You can very quickly put results on the table it is not a long development life cycle and such and the business guys love it"* (H6).

### 4.3   Challenges of Visual Analytics

**User Acceptance and Adoption.** Those interviewed generally agreed that the most pressing challenge they faced within their organisations was gaining user acceptance and adoption of VA. Many business users were familiar with tables that contained numbers and were reluctant to move to the visualisations presented to them.

B5 stated *"adoption can be a challenge especially in terms of the data, you need to, it needs to be as close to 100% as possible for people to trust what they are looking at."*

H2 commented that users were *"afraid that it may uncover information that they do not want other people to see"* and therefore *"some business units are unwilling to accept visual analytics."*

*"some of the business users make use of it and then there is some of them that don't that stick to their old way of doing things. So there is a lot of change management in getting people to use the tool. And that's where the challenge is"* (H6).

In order to encourage better user acceptance and adoption of VA, companies employed different methods. H2 said that their strategy to gain user acceptance and adoption *"was to remove their previous pdf generation and only give them the information they need visually. You know that people would revert to things that they know well if this meant that we had both the visualisation and the pdf. So by eliminating the pdf we have a higher adoption rate. But the users have also found, through this, that the visualisations hold a lot of value."*

Participants felt that if an effective change management strategy was enforced within an organisation, user acceptance and adoption would be less challenging.

**Lack of Effective Change Management:**  This was regularly discussed throughout the interviews. Two different aspects of change management were mentioned: numbers to visualisations; and mind-set change.

*Numbers to visualisations:*  Getting business users to use visualisations available to them instead of reverting to their table with numbers was a key aspect of change management that the researcher identified. Even though VA was available to various business users in different forms, such as dashboards, and visual analytic portals, users tended to revert to what they were familiar with. Many business users were not aware of the value or relevance that VA would deliver to their job functions.

H3 stated that: *"financial based people run the company and numbers are what they are used to and it is quite difficult to break through, they see a graph and it is difficult for them to see the number. They can see the spike but they want the number under it. It is change management actually."*

H4 said that *"Users will need to be trained in change management in order to move from A to B"*

B3 commented that: *"business users are used to seeing numbers, so that is a huge challenge to get them to think visually instead of number wise."*

B4 said: *"people prefer numbers so it is difficult to move them to visual images."*

B5 claimed that change management would involve *"sitting with various stakeholders and explaining to them how it is relevant to them. I think relevance is key. If people are not sure how it affects them they won't use it. Making them very aware that it will affect them and that they can rely on the data."*

*Mind-set change:*  There is also a need for business users to adopt a different mind-set when analysing visualisations. Many business users lacked the correct mind-set to analyse visual forms of data, and those that were familiar with numbers felt that the visuals did not give enough information. H5 stated that *"end users are close to the numbers and are not using VA as they think it gets in the way, some people feel that it detracts from the actual value. When people are exploring data or are unaccustomed to data they use visualisations but otherwise they prefer grids with numbers"*

H3 stated that the problem will always be that *"users are reluctant to change."*

This confirmed that organisations need to have effective change management initiatives in order to effectively move business users from numbers to visuals, as well as make business users aware of the potential benefits of analysing visual representations of data.

**Lack of Management Support.** Interviewees believed that with management support, a clear strategy would arise for the organisation to move forward with VA, and stressed that if management was personally involved with VA they would be able to pass the behaviour down to the business users working under them.

B2 stated *"if line managers can understand how this data can impact performance for their team members then they can also run with it as well. So I would say that it's everyone but, also first and foremost the managers because if we expect someone to change their behaviour or change something, if their line manager doesn't buy into it and doesn't understand it then it's not going to happen."*

H6 commented that within his organisation *"need business, and a champion in business to push it"*, and B3 stated that it is important that there be *"management support and input."* B4 however said that VA was *"not driven from management but from the users saying they want and need it."* When asked what played a big role in the effective adoption of VA, B5 stated that senior management adoption was crucial to the success of VA within the company. Without proper management support for a VA strategy, organisations might struggle to get the correct business user acceptance.

**Lack of Skills.** A pressing challenge that affects implementation of VA in an organisation is the lack of skills. Some interviewees complained that the learning curve is so big that it takes them a while to derive value from the tools because they are unsure of what exactly they can do with the tools. There is also limited support given to business users to facilitate their interaction with the visual analytic tool.

B3 stated that *"Each application has a different way in which it can be used. We don't have the training and don't know how to use the tools."*

B5 said an issue was the *"lack of skills and knowledge"* around VA.

H5 commented that in order to make VA successful in their organisation there is a *"need for training of the users."*

To combat the fact that users lack skills to use visual analytic tools, H2 said their team ensured that the VA presented to the business user are easy to use, stating: *"developers try to make sure that the users can easily navigate by themselves by giving tips, etc."*

If there were effective change management procedures in place, involving user training then lack of skills would be less of a challenge. Developers of VA should also strive, like H2, to ensure that VA presented to users with little to no training, are easy to navigate and understand.

**Uncertain Data Quality.** In any aspect of business intelligence and analytics, this is a key area. Interviewees noted that for users to accept and trust VA the data had to be accurate. B5 commented that *"data accuracy is important"* and *"adoption can be a challenge especially in terms of the data, you need to, it needs to be as close to 100% as possible for people to trust what they are looking at"*.

H2 stated that *"data quality is always an issue for us."*

H4 identified that the challenge was *"cleaning the data"* because *"if the underlying data model is wrong then people will get the wrong information, therefore data models need to be cleaned."*

Each organisation employs their own data storage technologies and has their own way of handling data quality. Lack of data quality could also be a result of a number of different aspects, such as lack of controls or poor database structure.

# 5 Discussion

Results from both the on-line survey and the interviews indicate that a number of South African business organisations are obtaining good benefits from VA, and that VA is being used across different levels and stakeholder roles. While value is being obtained, there are still significant barriers to more widespread use of VA. Having looked at the main themes and sub-themes that arose in interviews, we suggest how these may be acted upon to address challenges and improve benefits.

## 5.1 Good Practices for Visual Analytics

**Visual Analytics Strategy and Top Management Support.** The most effective VA programmes are when top management supports incorporating VA into their corporate strategy. This then needs to be communicated to all stakeholders to ensure that VA can be accepted as a "way forward" for the company. If this cannot be achieved, it is still most important to have a defined strategy for VA, and a high level champion in business to push it. If top management is not driving VA within the company, business users are less likely to engage with it.

**Stakeholder Involvement.** For VA to be successful and for users to accept the new tools more easily, the individuals or teams developing these tools should ensure stakeholder involvement during the development process. There should be regular and strong communication between the developers and business users about business processes and the visualisations being created.

**Effective Change Management.** This could help to overcome challenges of user acceptance and adoption as well as lack of skills. Proper change management would involve ensuring that all stakeholders are aware of the move from tables of numbers to VA, and how it could be used within the organisation. If this involved effective training and motivation of relevant stakeholders then they would not lack skills, and would be more likely to accept and apply VA within their job functions.

**Visual Analytic Tool Selection.** Organisations need to ensure that tools are able to fully integrate into the organisation in terms of database storage technology and requirements analysis. One tool may not necessarily cater for all business units within an organisation and multiple tools may be needed. A thorough proof of concept can establish how beneficial a VA tool will be to the organisation, and help satisfy return on investment (ROI) criteria.

**Interaction and Self Service.** Organisations that provide self-service to the users and allow them to interact with the data will cultivate a knowledge-seeking attitude within business users. This can generate enthusiasm and cultivate innovation within the organisation. Therefore, providing a VA tool that supports interaction and self-service is a key aspect to ensuring successful and widely used VA.

**Role-Based Visual Analytics.** Participants stressed that to encourage acceptance of VA, organisations should ensure that what was presented to users is role-based. Business users are reluctant to look at things that are not relevant to them, and therefore role-based presentation of data is important to ensure that the relevant information gets to the relevant level and type of employee.

**Data Management Strategy.** A clear data management strategy, with controls to ensure that data stored is of a high quality, and is well communicated to all stakeholders, will improve trust in the data and decision-making from VA. Data quality initiatives need to extend to unstructured and semi-structured data.

**Exception Highlighting.** Individuals or teams developing VA should create the visualisations in a way that highlights exceptions to the business user. This facilitates business users spotting when something is happening that should not happen, and will hopefully push them to try and understand what is wrong, and why. Exception highlighting is good for root cause analysis, and of course for also spotting "positive" anomalies and opportunities.

# 6   Conclusion and Recommendations

This study aimed to discover the value that organisations obtain from implementing VA and to obtain a better understanding of how VA is used within business. In doing so, both quantitative survey data and qualitative interviews were analysed. Although the smaller South African survey sample was purposive and not intended to be representative, comparative results obtained did not differ notably from those of the larger international survey by TDWI [9]. This suggests that many of the benefits and challenges uncovered may be applicable to a wide set of organisations internationally.

Respondents strongly recognized the extra value of VA over conventional numeric and tabular data. The benefits obtained from VA are both tangible and intangible, from changes in user behaviour, and encouragement of innovation, to better, faster decisions being made for the organisation. These benefits can be realised if the business takes the correct steps in integrating VA across technological, and also managerial and organisational aspects.

Organisations each face a different set of challenges but many of these are similar in nature. Challenges highlighted in this study were mostly organizational, rather than technology-based, as indicated in the literature reviewed. Broad use of VA is problematic without senior management support and strategic underpinning. The relevance and benefits of a change to VA from historic numeric presentation formats need to be well communicated, with adequate change management and skills training. At the same time, many employees with low affinity for numerical calculations may be converted to analytics and data-driven decision-making by an appropriate visual presentation format.

Some good practices are suggested for practitioners of VA that could help make VA successful and more widely used within an organization. Further research could be undertaken with a larger sample, could include in-depth and longitudinal case studies, or be focused on experience with VA usage in specific industry sectors.

**Acknowledgements.** This work is based on research partly supported by the South African National Research Foundation.

# References

1. Derksen, B., Luftman, J.: Key European IT management trends for 2016. CIONET Europe (2016)
2. Kappelman, L., McLean, E., Johnson, V., Torres, R.: The 2015 SIM IT issues and trends study. MIS Q. Exec. **15**(1), 55–83 (2016)
3. Gartner: Building the digital platform: insights from the 2016 gartner CIO agenda report (2016)
4. BARC: The BI survey 15. Business Application Research Centre, BARC GmbH (2016)
5. Howson, C.: Successful Business Intelligence, 2nd edn. McGraw Hill Education, New York (2014)
6. Negash, S.: Business intelligence. Commun. Assoc. Inf. Syst. **13**, 177–195 (2004)
7. Chen, M., Ebert, D., Hagen, H.: Data, information, and knowledge in visualization. IEEE Comput. Graph. Appl. **29**(1), 12–19 (2009)
8. Keim, D., Mansmann, F., Schneidewind, J., Ziegler, H.: Challenges in visual data analysis. In: Proceedings of 10th International Conference on Information Visualisation (IV 2006), pp. 9–16. IEEE (2006)
9. Stodder, D.: Data visualization and discovery for better business decisions: TDWI best practices report, 3rd Quarter, 2013. The Data Warehousing Institute (2013). www.tdwi.org
10. Yeh, R.: Visualization techniques for data mining in business context: a comparative analysis. In: Proceedings from Decision Science Institute, pp. 310–320 (2006). http://swdsi.org/swdsi06/Proceedings06/Papers/KMS04.pdf
11. Airinei, D., Homocianu, D.: Data visualization in business intelligence. In: Proceedings of 11th WSEAS International Conference on Mathematics and Computers in Business and Economics (MCBE), pp. 164–167 (2009)
12. Keim, D., Andrienko, G., Fekete, J.-D., Görg, C., Kohlhammer, J., Melançon, G.: Visual analytics: definition, process, and challenges. In: Kerren, A., Stasko, J.T., Fekete, J.-D., North, C. (eds.) Information Visualization. LNCS, vol. 4950, pp. 154–175. Springer, Heidelberg (2008). doi:10.1007/978-3-540-70956-5_7
13. Thomas, J., Cook, K.: Illuminating the path: the research and development agenda for visual analytics. IEEE Computer Society Press (2005)

14. Williams, B.G., Boland, R.J. Jr., Lyytinen, K.: Shaping problems, not decisions: when decision makers leverage visual analytics. In: Proceedings of 21st AMCIS Conference, pp. 1–15 (2015)
15. Davenport, T., Harris, J.: Competing on Analytics: The New Science of Winning. Harvard Business School Press, Boston (2007)
16. Fisher, D., DeLine, R., Czerwinski, M., Drucker, S.: Interactions with big data analytics. Interactions **19**(3), 50–59 (2012)
17. Ransbotham, S., Kiron, D., Prentice, P.K.: Beyond the hype: the hard work behind analytics success. MIT Sloan Manag. Rev., March 2016
18. Sabherwal, R., Becerra-Fernandez, I.: Business Intelligence Practices, Technologies and Management. Wiley, Hoboken (2011)
19. Bačić, D., Fadlalla, A.: Business information visualization intellectual contributions: an integrative framework of visualization capabilities and dimensions of visual intelligence. Decis. Support Syst. **89**, 77–86 (2016)
20. Zhang, L., Stoffel, A., Behrisch, M., Mittelstadt, S., Schreck, T., Pompl, R., Weber, S., Last, H., Keim, D.: Visual analytics for the big data era—a comparative review of state-of-the-art commercial systems. In: IEEE Symposium on Visual Analytics Science and Technology, pp. 173–182 (2012)
21. Heer, J., Agrawala, M.: Design considerations for collaborative visual analytics. Inf. Vis. **7**(1), 49–62 (2008)
22. Marjanovic, O.: Using practitioner stories to design learning experiences in visual analytics. In: 2014 IAIM Proceedings, Paper 4, pp. 1–13 (2014)
23. Fisher, B., Green, T.M., Arias-Hernandez, R.: Visual analytics as a translational cognitive science. Topics Cogn. Sci. **3**, 609–625 (2011)
24. Cybulski, J.L., Keller, S., Nguyen, L., Saundage, D.: Creative problem solving in digital space using visual analytics. Comput. Hum. Behav. **42**, 20–35 (2015)
25. Keim, D.A., Mansmann, F., Schneidewind, J., Thomas, J., Ziegler, H.: Visual analytics: scope and challenges. In: Simoff, S.J., Böhlen, M.H., Mazeika, A. (eds.) Visual Data Mining. LNCS, vol. 4404, pp. 76–90. Springer, Heidelberg (2008). doi:10.1007/978-3-540-71080-6_6
26. Ribarsky, W., Wang, D.X., Dou, W.: Social media analytics for competitive advantage. Comput. Graph. **38**, 328–331 (2014)
27. Laramee, R.S., Kosara, R.: Challenges and unsolved problems. In: Kerren, A., Ebert, A., Meyer, J. (eds.) Human-Centered Visualization Environments. LNCS, vol. 4417, pp. 231–254. Springer, Heidelberg (2007). doi:10.1007/978-3-540-71949-6_5
28. Kohlhammer, J., Keim, D., Pohl, M., Santucci, G., Andrienko, G.: Solving problems with visual analytics. In: European Future Technologies Conference and Exhibition, vol. 7, pp. 117–120 (2011)
29. Thomas, J., Kielman, J.: Challenges for visual analytics. Inf. Vis. **8**(4), 309–314 (2009)
30. Wong, P.C., Shen, H.-W., Johnson, C.R., Chen, C., Ross, R.B.: The top 10 challenges in extreme-scale visual analytics. IEEE Comput. Graph. Appl. **32**(4), 63–67 (2014)
31. Saunders, M., Lewis, P., Thornhill, A.: Research Methods for Business Students, 5th edn. Pearson Education/Prentice Hall, London (2009)
32. Bijker, M., Hart, M.: Factors influencing pervasiveness of organisational business intelligence. In: 3rd International Conference on Business Intelligence and Technology, BUSTECH 2013, pp. 21–26 (2013)

33. DePietro, R., Wiarda, E., Fleischer, M.: The context for change: organization, technology and environment. In: Tornatzky, L.G., Fleischer, M. (eds.) The Processes of Technological Innovation, pp. 151–175. Lexington Books, Lexington (1990)
34. Braun, V., Clarke, V.: Using thematic analysis in psychology. Qual. Res. Psychol. **3**(1), 77–101 (2006)
35. Guest, G., MacQueen, K.M., Namey, E.E.: Applied Thematic Analysis. Sage, Thousand Oaks (2011)

# Analytics and Decision

# Conceiving Hybrid What-If Scenarios Based on Usage Preferences

Mariana Carvalho and Orlando Belo[✉]

ALGORITMI R&D Center, Department of Informatics, School of Engineering,
University of Minho, 4710-057 Braga, Portugal
obelo@di.uminho.pt

**Abstract.** Nowadays, enterprise managers involved with decision-making processes struggle with numerous problems related to market position or business reputation of their companies. Owning the right and high quality set of information is a crucial factor for developing business activities and gaining competitive advantages on business arenas. However, today retrieving information is not enough anymore. The possibility to simulate hypothetical scenarios without harming the business using What-If analysis tools and to retrieve highly refined information is an interesting way for achieving such business advantages. In a previous work, we introduced a hybridization model that combines What-If analysis and OLAP usage preferences, which helps filter the information and meet the users' needs and business requirements without losing data quality. The main advantage is to provide the user with a way to overcome the difficulties that arise when dealing with the conventional What-If analysis scenario process. In this paper, we show an application of this methodology using a sample database, and compare the results of a conventional What-if process and our methodology. We designed and developed a specific piece of software, which aims to discover the best recommendations for What-If analysis scenarios' parameters using OLAP usage preferences, which incorporates user experience in the definition and analysis of a target decision-making scenario.

**Keywords:** What-If analysis · On-Line analytical processing · Usage preferences · Analysis systems specification · Multidimensional databases

## 1 Introduction

Nowadays, decision-making processes can be improved with the use of analytical tools and data mining techniques and models. The presence of analytical information systems and the availability of techniques and models for multidimensional data exploration and analysis within a company is no longer a novelty in enterprise business environments. In addition, there has been a noticeable increase in the number and quality of data retrieving and handling processes created, developed or used by companies n their business processes. To compete in a knowledge-based society, and to be a part of a rapidly changing global economy, companies must try gaining some competitive advantage from a better use of information and knowledge.

© Springer International Publishing AG 2017
I. Linden et al. (Eds.): ICDSST 2017, LNBIP 282, pp. 119–132, 2017.
DOI: 10.1007/978-3-319-57487-5_9

What-If analysis [9] is one way to gain some competitive advantage effectively. What-If analysis processes allow for creating simulation models, aiming to explore the behavior of complex business systems. More pragmatically, they help analyzing the effects on the behavior of a business system caused by the change of variables and values. These changes usually cannot otherwise be discovered by historical data manual analysis process. The main advantage of creating a simulation model is to implement changes on model business' variables without endangering the real business [17]. Decision-makers can create What-If scenarios to test and validate their business hypothesis and support their decisions, being the safer solution to address any doubt and to ensure that the subsequent decision will be succeeded. *Online Analytical Processing* (OLAP) [13] systems are one of the most predominant tools for decision-support systems. They provide means for business analytics as well as multidimensional views over business data, which are very efficient ways for analyzing businesses activities and organizations. Decision-makers usually run complex queries in OLAP systems, which may return huge volumes of data and may be quite difficult to analyze. Thus, it is essential to filter this information in a way that data do not lose significance, being adjusted according to the users' needs and the business requirements. The extraction of OLAP usage preferences according to each analytic session promoted by a user may come as an advantage to decision-makers, since it provides a very effective way to personalize analytical sessions and multidimensional data structures, acting as their decision-making support [10, 14]. One of the pitfalls of a What-If analysis is the lack of expertise of the user in its design and implementation. Often the user is not familiar with the business and may not choose the most correct parameters in the scenario, providing an outcome that probably is not the most adequate. Therefore, we developed a hybridization process [5], which consists in the integration of the OLAP usage preferences in a What-If scenario process. With this model, we intend to improve the quality and effectiveness of simulation models and to provide the user a way to overcome the difficulties that usually arise when dealing with conventional What-If scenario process.

In this paper, we refer to the methodology and present an application example using a sample data case that illustrates how the use of preferences in What-if analysis can significantly improve the results when compared with conventional What-if analysis processes. This paper is organized as follows. In Sect. 2 we present an overview about preferences and the What-If process and how they are used in decision support decision related areas. Next in Sect. 3, we show how the conventional What-If analysis works. In Sect. 4 we present an example of application of our methodology, explaining all the steps of the hybridization process since the extraction of the rules until the creation of the What-If scenario. Finally, in Sect. 5, we conclude the paper and discuss some possible future research directions.

## 2    Related Work

The methodology we designed introduces preferences in a conventional What-If process application scenario. We show how preferences are used to improve decision-making process, either in relational databases, in OLAP environments, or even in daily tasks.

The research on databases preferences goes back [20], which was one of the first works that presented and discussed a preference mechanism as an extension of a query language. Later, in [2] a formal framework was proposed for expressing and combining user preferences to address the problem of the high quantity of available on-line information. Furthermore, the authors of [6] presented a logical framework for formulating preferences and embedding them into relational query languages, which not imposing any restrictions on the preference relations, and allowing for arbitrary operations and predicate signatures in preference formulas. After this, a different approach to database preferences queries was presented in [11], presenting and discussing the way we can deal with preferences in a logic manner using symbolic weights. At the same time, other new approaches emerged, and new applications areas arose as well. See, for example, the work presented in [21] about traffic analysis, where authors presented a set of methods for including driver preferences and time-variant traffic condition in route planning.

Nowadays, What-If analysis is used in several areas, mostly for improving the performance of tools. In [8] it was used some dedicated provisioned autonomous representations (PARs) to improve a tool (Caravan) that allows for performing What-If analysis. The authors of [16] developed a tool to interactively model and run generic What-If analysis to help measuring the success of companies. In [4] it is presented a method for estimating crash and injury risks from off-road glance behavior for crashes using What-If simulation, and in [7] we find a way to combine the Structural Equation Modeling (SEM) results with Fuzzy Cognitive Map (FCM) to support What-If analysis and get better results. More recently, the work presented in [23] revealed a tool-supported What-If based approach to detect and avoid inconsistencies in database schemas. Additionally, in [26] we can find the reference of a What-If based tool to analyze and improve the results given by ShipitSmarter, while in [15] we discovered a methodology that can estimate the distribution of cloud latency of a request under a What-If scenario. Finally, [22] shows us how to improve an optimization model and decision support tool already developed, including adding What-If capability in order to help MoDOT to identify the impact of changing policies, and [3] presents an improved approach of the VATE2 system by introducing What-If analysis.

## 3  Conventional What-If Analysis

In this project we used a set of data objects extracted from "AdventureWorksDW2014" [24], a small data warehouse provided by Microsoft as a product sample for Microsoft SQL Server. It uses the sample AdventureWorks OLTP (Microsoft SQL Server Product Samples: Database, 2015) database as its data source. The AdventureWorks database contains information about a fictitious, multinational manufacturing company, including very interesting data objects like employees, stores, production, sales and customers (resellers or individuals). We carried out our analysis on the profit sales of the company. Basically, the goal is to improve profit sales, collecting more resellers. This profit source is the company's strong profit source. In Fig. 1, we can see the difference between sales amount of customers: individuals in the left and resellers in the right.

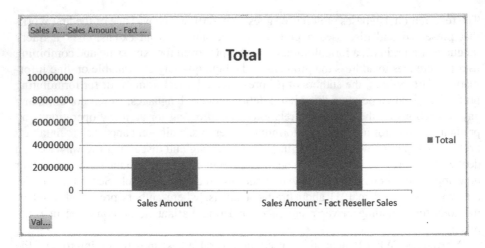

**Fig. 1.** Sales amount by customers (individuals and resellers).

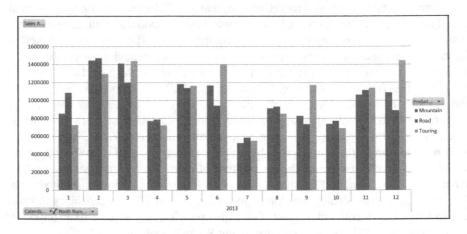

**Fig. 2.** A conventional business analysis scenario. (Color figure online)

We want to focus on reseller's characteristics; for example, we want to focus on product lines that resellers seek most: "What if we want to increase the profit sales by 10%, focusing mainly product lines acquired by resellers in 2013?" To do this, we intend to discover which attributes in the data are related to the goal attribute ("Product Line") in order to add new information to the conventional What-If scenario (Fig. 3). With this analysis we can see which product lines are the most required by resellers and how much money the company made with each product line in 2013. Without OLAP preferences, an analyst would select for his application scenario the attributes "Sales Amount" represented by the Y axis, with a range of 0 to 1 600 000; and represented by the X axis: "Calendar Year" (2013), "Month Number Of Year" with a range of 1 to 12 which represents the months of a year, and "Product Line" which can be "Mountain" in blue, "Road" in red and "Touring" in green. If so, we got a chart like the one presented in Fig. 2. Considering our

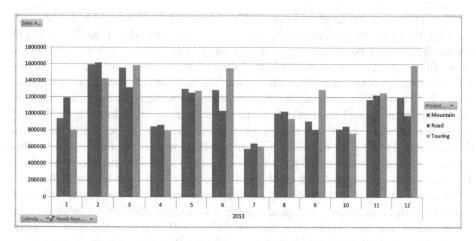

**Fig. 3.** A What-If analysis scenario without preferences. (Color figure online)

what-if question (or goal analysis) and following a conventional What-If methodology, we get the scenario represented in Fig. 3.

With these two charts we can pass on potential knowledge for decision making to managers, such as: "Touring" line is the most acquired by resellers and it is the product line that earn more money, especially in March, June, September, November and December, followed by "Mountain", which is the best seller in May. And finally, "Road", which is the product line that earns less money – however, this product line earns more money in January, February, April, July, August and October. This company earns less money in April, July and October, will be helpful increase sales in these months. Managers can decide to improve product for a certain product line, in order to generate more profit. This information is viable, but is very generic and abstract, and better decisions can be reached with more detailed information. Proposing a new methodology, we aim to show that What-If scenarios based on preferences are richer the other scenarios that would normally be obtained, as happened in Fig. 3. Then, with this extra information (perspectives of analysis), a What-If scenario would be more accurate and specific, leading to better results. For instance, the extraction of association rules may show that the product line is strongly related with other resellers' characteristics.

## 4  What-If Analysis Hybridization

Previously, in [5] it was presented a methodology for approaching hybrid models based on preferences for What-If scenarios. The process of extracting OLAP usage preferences and use them to enrich What-If scenarios, considers five distinct phases:

1. Selecting the data view.
2. Constructing the cube.
3. Extracting association rules.
4. Extracting association rules usage preferences.
5. Performing What-If analysis using preferences.

The application we developed allows for the user: (i) to create What-If scenarios choosing the available attributes of his choice (conventional What-If analysis); (ii) to consult the mining models' item sets and association rules; (iii) to combine both options (i) and (ii) together, which we call the hybrid model that aims to create What-If scenarios using preferences obtained with the mining models' association rules. It provides users with information about association rules that were extracted from the cube structure, creates preferences, and recommends them to the user, in order to create a What-If application scenario. In the first step of this process, the user sets values to filter both the support and probability of both item set and rules (as was seen before in step 1). This way, it is possible to refine users' preferences, leading to a specific and filtered outcome. The association rules extracted from the mining model can also be filtered as the item sets (as seen in step 3 of the process) and displayed ordered by decreasing values of probability. In a later phase, the application suggests a set of item sets (contained in the chosen rules). The user chooses the item sets, which will be part of the What-If application scenario. In the example introduced earlier in this paper for the conventional What-if analysis, we show an application and example of our methodology. Following the methodology we designed, we begin with a view selection process over the Data Warehouse [19], starting with a small case study, which considers a single fact table "FactResellerSales" – and related dimension tables – detailed information about Resellers' companies ("Dim Reseller"), product's information ("Dim Product"), the calendar ("Dim Date"), employee's information ("Dim Employee"), information about current and past promotions ("Dim Promotion"), information about standard ISO currencies ("Dim Currency") and the list of sales territories ("Dim Sales Territory"). These tables are the most adequate to support our goal analysis, once this fact table contains the information about resellers sales' details.

In the second phase, we create and analyze a specific data cube [13]. The data cube is a multidimensional database, in which each cell within the cube structure contains numeric facts called measures that are categorized by dimensions. The cube created using the previous data objects is a simple example to illustrate the methodology. In the next phase we perform On-Line Analytical Mining (also called OLAP mining), which consists in apply a mining technique to the OLAP Cube. In our example, an Association Rule algorithm is the most adequate mining technique to extract the preferences OLAP cube. We create the mining structure and define the mining model to support a mining association process that runs in the third phase over the cube we created [12]. To do that we selected the Microsoft Association Rules algorithm [1] that comes with Microsoft Analysis Services. This Apriori based algorithm fits well on mining processes that involves recommendation engines or processes for finding correlations between different attributes in a given dataset – in our case we have a recommendation engine for the suggestion of the items that are most likely to appear together in a particular search of a What-If scenario. As other Apriori based association algorithms, we can define the minimum and the maximum support values to control the number of item sets that are generated, and we can also restrict the number of rules that a model produces by setting a value for a minimum probability. After getting the list of association rules, it is necessary to filter this list in order to extract the OLAP preferences and suggest them to the user. Now it is time to define formally what is a preference, introducing it with a simple working example based on the works presented in [18, 25]. Thus, given a set of attributes A, a preference P is a strict partial order defined as P (A, <P), where <P is an irreflexive, transitive and asymmetric binary relation <P ⊆ dom(A) × dom(A). If X <P Y, then 'Y is preferred to X'. A preference P = (A, <P) is an irreflexive, transitive and asymmetric binary relation <P on the domain of values of attributes set A. Let see how this works. If we want to analyze how sales vary with product line acquired by resellers, to set its preferences a user need to choose one of the elements included in the set of the frequent item sets. This will allows for choosing the rules that will be used to set user preferences. Thus, assuming that the user chooses "Product Line", the attribute "Product Line" is preferred to the attribute "Geography Key", "Annual Sales", "Last Order Year", "Business Type", "First Order Year", and so on. Thus, "Geography Key" is equivalent to "Annual Sales", Annual Sales" is equivalent to "Last Order Year", "Last Order Year" is equivalent to "Business Type", and so on. Based on this set of previous preferences, it is possible to select a set of association rules that contains the attribute "Product Line" (Fig. 4).

| |
|---|
| Geography Key = 138 - 279, Product Line = Mountain -> Annual Sales >= 1500000 |
| Last Order Year < 2011, Product Line = Road -> Business Type = Specialty Bike Shop |
| First Order Year < 2010, Product Line = Road -> Order Frequency = A |
| Bank Name = International Security, Year Opened >= 1999 -> Product Line = Mountain |
| Last Order Year < 2011, Product Line = Road -> Order Frequency = A |
| First Order Year < 2010, Product Line = Road -> Business Type = Specialty Bike Shop |
| Geography Key = 279 - 406, Min Payment Amount = 100.0790444416 - 800 -> Product Line = Road |
| Product Line = Touring, Year Opened >= 1999 -> Number Employees = 19 - 39 |
| Year Opened < 1980, Business Type = Specialty Bike Shop -> Product Line = Road |
| Year Opened < 1980, Order Frequency = A -> Product Line = Road |
| First Order Year = 2010 - 2011, Last Order Year >= 2012 -> Product Line = Road |
| Year Opened = 1980 - 1986, Geography Key = 406 - 523 -> Product Line = Road |
| Last Order Year = 2011 - 2012, Min Payment Amount < 100.0790444416 -> Product Line = Road |
| Number Employees = 19 - 39, Geography Key >= 523 -> Product Line = Road |
| Year Opened = 1980 - 1986, Min Payment Amount = 100.0790444416 - 800 -> Product Line = Road |

**Fig. 4.** Association Rules including the "Product Line" attribute.

Accordingly its own business preferences, the user may choose N association rules, for example the top 3 association rules (Fig. 5) of the previous set (Fig. 4), which will be used later to define his OLAP preferences. This means that the item sets contained in the filtered association rules will be suggested to the user as preferences. For example, if the returned list of association rules is the list in Fig. 5, the recommendations to the user will are "Product Line", obviously, "Geography Key", "Annual Sales", "Last Order Year", " Business Type", " First Order Year and Order Frequency". After this step, the user chooses within the recommendations, the item sets of his preference that will be used as configuration parameters in the What-If scenario (Fig. 6).

| Geography Key = 138 - 279, Product Line = Mountain -> Annual Sales >= 1500000 |
| Last Order Year < 2011, Product Line = Road -> Business Type = Specialty Bike Shop |
| First Order Year < 2010, Product Line = Road -> Order Frequency = A |

**Fig. 5.** Top 3 Association Rules with "Product Line".

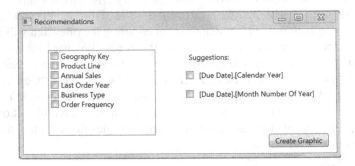

**Fig. 6.** UI Recommendations.

This last phase proceeds as is shown in Fig. 7. Firstly, we use an OLAP data cube as input. The input data will be used to define the application scenario based on historical data extract from previous OLAP sessions. Then, we use this data to define a simulation model. A simulation model is the main focus of a What-If application. Commonly, this model is a representation of a real business model and usually is composed into several application scenarios. Each scenario considers a set of business variables (the source variables) and a set of setting parameters (scenario parameters). It is the user responsibility to delineate the axis of analysis, the set of values for analyzing, and the set of values to change according to the goals defined previously. Then, the What-If process is performed with an appropriate tool. To run a simulation model, which is a scenario based on historical data, it is required to have a tool that can perform What-If scenario analysis, in order to get a prediction scenario. The What-If analysis tool calculates and lets the user to explore and analyze the impact of the changes in the setting values of the entire application scenario. It is the user who is responsible for accepting the new data cube or returning to change the settings of the application scenario and make the changes required over to the target data.

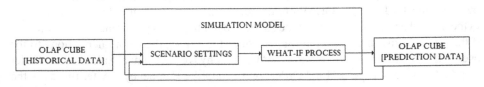

**Fig. 7.** An overview of a What-If analysis process.

Now we present an example of the business application. To support and perform What-If analysis processes we choose Microsoft Office Excel, since it allows for creating

PivotTable reports based on OLAP source data. OLAP PivotTable Extensions is an Excel add-in, which extends the functionality of PivotTables on Microsoft Analysis Services multidimensional structures. We show in Fig. 8 an example of the business application scenario containing information about "Sales Amount" in 2013, with axis of analysis "Sales Territory Group", "Business Type" and "Product Line". In case of example, the following scenarios created by Excel contain the attribute "Sales Territory Group", instead of "Geography Key" in order to be easy to understand the What-If charts.

| | A | B | C | D | E | F | G | H |
|---|---|---|---|---|---|---|---|---|
| 1 | Sales Amount | | | | Product Line ▾ | | | |
| 2 | Calendar Year ▾ | Month Number Of Yea ▾ | Sales Territory Group ▾ | Business Type ▾ | Mountain | Road | Touring | Total Geral * |
| 3 | 2013 | 1 | Europe | Specialty Bike Shop | | 8475,3042 | 15433,44517 | 23908,74937 |
| 4 | | | | Value Added Reseller | 35089,28368 | 44855,8638 | 24959,49423 | 104904,6417 |
| 5 | | | | Warehouse | 62679,48214 | 92748,01382 | 142548,629 | 297976,1249 |
| 6 | | | Europe Total * | | 97768,76582 | 146079,1818 | 182941,5684 | 426789,516 |
| 7 | | | North America | Specialty Bike Shop | 19673,88192 | 61334,15739 | 28078,5626 | 109086,6019 |
| 8 | | | | Value Added Reseller | 436726,1289 | 273936,4271 | 132847,5684 | 843510,1243 |
| 9 | | | | Warehouse | 385332,1295 | 711185,7166 | 401504,3195 | 1498022,166 |
| 10 | | | North America Total * | | 841732,1403 | 1046456,301 | 562430,4505 | 2450618,892 |
| 11 | | | Pacific | Specialty Bike Shop | | | 21648,24794 | 21648,24794 |
| 12 | | | | Value Added Reseller | | | 30641,35976 | 30641,35976 |
| 13 | | | | Warehouse | | | 2517,57792 | 2517,57792 |
| 14 | | | Pacific Total * | | | | 54807,18562 | 54807,18562 |
| 15 | | 1 Total * | | | 939500,9061 | 1192535,483 | 800179,2044 | 2932215,593 |
| 16 | | 2 | Europe | Specialty Bike Shop | 26613,5562 | 18186,81205 | 58829,64549 | 103630,0137 |
| 17 | | | | Value Added Reseller | 44102,75639 | 78721,1568 | 136673,1356 | 259497,0488 |
| 18 | | | | Warehouse | 305000,7445 | 94569,24422 | 526056,4057 | 925626,3944 |
| 19 | | | Europe Total * | | 375717,0571 | 191477,2131 | 721559,1868 | 1288753,457 |
| 20 | | | North America | Specialty Bike Shop | 60609,27356 | 107488,3519 | 46946,89197 | 215044,5174 |
| 21 | | | | Value Added Reseller | 605363,6743 | 788213,0313 | 106434,1659 | 1500010,871 |
| 22 | | | | Warehouse | 545305,3192 | 530446,461 | 406998,9541 | 1482750,734 |

**Fig. 8.** A business application scenario using a MS Excel PivotTable.

Excel can be used as an OLAP analytical tool to easily analyze and modify data stored on data cubes. It is possible to modify data using a PivotTable and to recalculate all data as necessary, and, if the outcome is acceptable, to publish all changes so that they are copied into the OLAP cube. It is this property of Excel that allows for performing What-If Analysis and to create new application scenarios with the recalculated data. After choosing the parameters for the What-If scenario, the user can make some changes – e.g. increasing the total sales values by 10%. Then, Excel calculates how the new value will modify the old values, based on the properties of 'What-If Analysis Settings'. Microsoft Excel allows for the user to calculate data with changes that were made manually (the user decides when the changes are made) or automatically (when each value is changed), to choose the allocation method ('Equal Allocation' or 'Weighted Allocation') and finally to select the value to allocate - the value entered is divided by the number of allocations or it is incremented based on the old value. Next, the new What-If scenario (the scenario with new calculated values) will be displayed to the user. We set a simple possible business scenario in case of the example in Fig. 9(i), (ii) and (iii). Instead of using the suggestions extracted from the association rules in order to be easier to understand the results and conclusions of the analysis of the scenario. The set of attributes used as scenario parameters is different, but as we will see, the advantages of using our methodology are still the same.

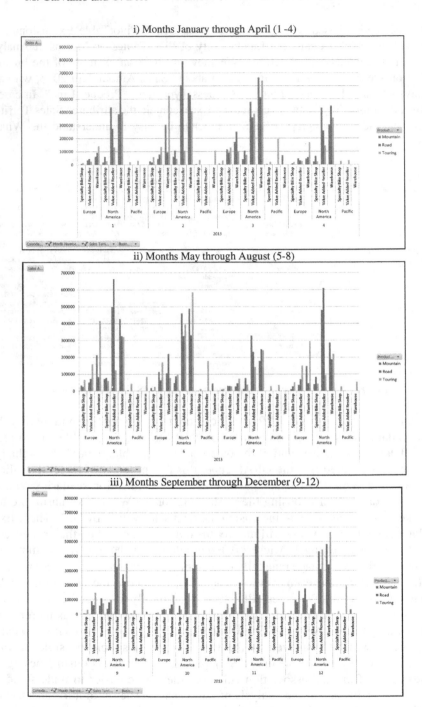

**Fig. 9.** The enhanced What-If scenario (Color figure online)

Considering the analysis scenario represented by the PivotTable in Fig. 8, with the set of attributes "Calendar Year" (2013), "Month Number of Year" with a range of 1 to 12, "Sales Territory Group" which can be "Europe", "North America" or "Pacific", "Business Type" which can be "Specialty Bike Shop", "Value Added Reseller" or "Warehouse", and "Product Line" which can be "Mountain", "Road" or "Touring"; all represented by the X axis; and "Sales Amount" with a range of 0 to 900 000, represented by the Y axis. After we perform the What-If process (recall that we intend to increase the final value of sales amount in 10%.), we get the graphic form represented in Fig. 9. Due to complexity of the graphics, we divide the year in three figures. With these three charts we discover more information and get more specific information about resellers. Therefore, separate analysis should be carried depending on the characteristics of the resellers. In this case, we can analyze how sales vary according to region, business type and month. We will analyze which spent more and less, and discover patterns, which cannot be otherwise discovered with conventional What-If analysis.

Overall, in 2013, resellers spent more money in products of the "Touring" product line (with sales amount over 13 800 000), followed by "Mountain" line (with sales amount over 13 000 000) and "Road" line (with sales amount over 12 800 000). This information was already discovered with the conventional What-If analysis in Fig. 3. But we can specify the group of resellers (region and business type) that acquired each one of the available product lines. The disparity of the sales amount values between spent by North America and the other two regions (Europe and Pacific); and between "Value Added Reseller", "Warehouse" and "Specialty Bike Shop" resellers is irrefutable: resellers of North America spent more money, followed by Europe and Pacific; and "Specialty Bike Shop" resellers spent less money in the available product lines.

In North America, "Value Added Reseller" resellers spent more money with the "Road" line (with sales amount over 600 000) in February, May, August and November, but in January "Warehouse" resellers spent over 700 000 in the same product line. "Specialty Bike Shop" resellers spent less money (sales amount under 20 000) in January, July and October with the "Mountain" line, but spent over 100 000 in the "Road" line in February and March. In Europe, "Warehouse" resellers spent more money (sales amount over 400 000) in the "Touring" line in February, May and November. "Specialty Bike Shop" resellers spent less money (sales amount under 5 000) in the "Road" line in March, June, September and December, but spent over 60 000 in "Touring" products in May and November. In Pacific, there are no records of sales of "Road" products. "Value Added Reseller" resellers spent more money with the "Touring" line (sales amount over 150 000) in March, June, September and December. And finally, "Specialty Bike Shop" resellers spent less money (sales amount under 1 000) in the "Mountain" products, but spend over 35 000 in "Touring" products in February, May and November.

Adding preferences to this process, suggestions are made to the analyst and the resulting What-If scenarios clearly contain more information, making easier to the analyst to discover and pass on potential knowledge to decision-makers. For illustrative purposes, in our example, considering the last What-If scenario represented in Fig. 9 and comparing it to the What-If scenario without preferences (Fig. 3), we can restrict even more the target audience and pass on more detailed information. A way to increase profit sales would be to improve products of the "Road" line in North America, and "Touring" products in Europe and Pacific, in order to lead resellers to invest in better products.

Combining What-If analysis with preferences improves significantly the potential knowledge that an analyst can extract from the scenarios when compared with a conventional What-If process. Due to the use of an association rules algorithm, we can discover the strongly related attributes to the main goal analysis. Analysts may not have any acquaintance about the business model and restrict themselves only to the What-If question content (What-if scenario without preferences). This must be avoided, because a user that has limited knowledge about the business domain or even about the What-If process to be implemented influences the entire process negatively, leading to poor results.

## 5    Conclusions and Future Work

In this paper we show how OLAP preferences can contribute for enhancing a What-If scenario, improving the quality and effectiveness of decision models where perceptions from the user point of view can make the difference in a decision-making process. We implemented a decision-support system with the ability to receive a What-If scenario that incorporates usage analytical preferences for improving the simulation of a given business application scenario. The system has the ability to provide its users with the most adequate scenario parameters according to their needs taking into consideration a set of OLAP preferences that were extracted from past OLAP sessions. This contributes significantly to enrich and to make more valuable a What-If scenario for a particular business domain. The design and implementation of our system can help in the evaluation of business scenarios that integrate process solutions for analytical data exploration environments.

Despite the several advantages of using preferences, there are some drawbacks related to this process. We can face some difficulties in the What-If process. In a first stage of the What-If process, if the goal analysis is not done correctly, What-If questions and scenarios will be not correctly defined or the preferences outcome will be not reliable. Thereafter, performed What-If process will be not the most suitable process and thus the obtained prediction will be different of what is expected as a normal behavior of a real business system. Finally, this whole application process depends strongly on the user. In order to overcome this kind of drawbacks, we mainly aim at restructuring automatically the What-If scenarios, discarding the user's dependency and finding a way of overcoming the limitation we found in some Excel functions.

**Acknowledgments.**    This work has been supported by COMPETE: POCI-01-0145-FEDER-007043 and FCT - Fundação para a Ciência e Tecnologia within the Project Scope: UID/CEC/00319/2013.

# References

1. Agrawal, R., Srikant, R.: Fast algorithms for mining association rules. In: Proceeding 20th International Conference Very Large Data Bases, VLDB, vol. 1215, pp. 487–499 (1994)
2. Agrawal, R., Wimmers E.: A framework for expressing and combining preferences. ACM SIGMOD Record **29**(2) (2000)
3. Angelini, M., Ferro, N., Santucci, G., Silvello, G.: A visual analytics approach for what-if analysis of information retrieval systems. In: Proceeding 39th Annual International ACM SIGIR Conference on Research and Development in Information Retrieval (SIGIR 2016). ACM Press, New York, USA (2016)
4. Bärgman, J., Lisovskaja, V., Victor, T., Flannagan, C., Dozza, M.: How does glance behavior influence crash and injury risk? A 'what-if' counterfactual simulation using crashes and near-crashes from SHRP2. Transp. Res. Part F: Traffic Psychol. Behav. **35**, 152–169 (2015)
5. Carvalho, M., Belo, O.: Enriching what-if scenarios with OLAP usage preferences. In: 8th International Joint Conference on Knowledge Discovery, Knowledge Engineering and Knowledge Management, vol. 1, pp. 213–220 (2016)
6. Chomicki, J.: Preference formulas in relational queries. ACM Trans. Database Syst. (TODS) **28**(4), 427–466 (2003)
7. De Maio, C., Botti, A., Fenza, G., Loia, V., Tommasetti, A., Troisi, O. Vesci, M.: What-if analysis combining fuzzy cognitive map and structural equation modeling. In: 2015 Conference on Technologies and Applications of Artificial Intelligence (TAAI) pp. 89–96 (2015)
8. Deutch, D., Ives, Z. G., Milo, T., Tannen, V.: Caravan: provisioning for what-if analysis. In: CIDR (2013)
9. Golfarelli, M., Rizzi, S. Proli, A.: Designing what-if analysis: towards a methodology. In: DOLAP 2006, Arlington, Virginia, USA, pp. 51–58 (2006)
10. Golfarelli, M., Rizzi, S.: Expressing OLAP preferences. In: Winslett, M. (ed.) SSDBM 2009. LNCS, vol. 5566, pp. 83–91. Springer, Heidelberg (2009). doi:10.1007/978-3-642-02279-1_7
11. Hadjali, A., Kaci, S., Prade, H.: Database preferences queries – a possibilistic logic approach with symbolic priorities. In: Hartmann, S., Kern-Isberner, G. (eds.) FoIKS 2008. LNCS, vol. 4932, pp. 291–310. Springer, Heidelberg (2008). doi:10.1007/978-3-540-77684-0_20
12. Han, J.: OLAP mining: an integration of OLAP with data mining. In: Proceedings of the 7th IFIP, pp. 1–9 (1997)
13. Harinarayan, V., Rajaraman, A. Ullman, J.: Implementing data cubes efficiently. ACM SIGMOD Record. **25**(2) (1996)
14. Jerbi, H., Ravat, F., Teste, O., Zurfluh, G.: Preference-based recommendations for OLAP analysis. In: Pedersen, T.B., Mohania, Mukesh K., Tjoa, A.M. (eds.) DaWaK 2009. LNCS, vol. 5691, pp. 467–478. Springer, Heidelberg (2009). doi:10.1007/978-3-642-03730-6_37
15. Jiang, Y., Sivalingam, L. R., Nath, S., Govindan, R.: WebPerf: evaluating what-if scenarios for cloud-hosted web applications. In: Proceedings of the 2016 conference on ACM SIGCOMM 2016 Conference, pp. 258–271 (2016)
16. Klauck, S., Butzmann, L., Müller, S., Faust, M., Schwalb, D., Uflacker, M., Sinzig, W., Plattner, H.: Interactive, flexible, and generic what-if analyses using in-memory column stores. In: Renz, M., Shahabi, C., Zhou, X., Cheema, M.A. (eds.) DASFAA 2015. LNCS, vol. 9050, pp. 488–497. Springer, Cham (2015). doi:10.1007/978-3-319-18123-3_29
17. Kellner, M.I., Madachy, R., Raffo, D.: Software process simulation modeling: Why? What? How? J. Syst. Softw. **46**(2), 91–105 (1999)
18. Kießling, W.: Foundations of preferences in database systems. In: Proceedings of the 28th International Conference on Very Large Data Bases, VLDB Endowment, pp. 311–322 (2002)

19. Kimball, R., Ross, M.: The data warehouse toolkit: the complete guide to dimensional modeling. Wiley (2011)
20. Lacroix, M., Lavency P.: Preferences: putting more knowledge into queries. In: VLDB 1987 (1987)
21. Letchner, J., Krumm J., Horvitz E.: Trip router with individualized preferences (trip): incorporating personalization into route planning. In: Proceedings of the National Conference on Artificial Intelligence, vol. 21, no. 2, p. 1795 (2006)
22. McGarvey, R.G., Matisziw, T., Noble, J.S., Nemmers, C.J., Karakose, G., Krause, C.: Improving Striping Operations through System Optimization-Phase 2 (2016)
23. Meurice, L., Nagy, C., Cleve, A.: Detecting and preventing program inconsistencies under database schema evolution. In: IEEE International Conference Software Quality, Reliability and Security (QRS), pp. 262–273 (2016)
24. Microsoft SQL Server Product Samples: Database (2015). http://msftdbprodsamples.codeplex.com/, Accessed 13 Feb 2016
25. Ore, O., Ore, Y.: Theory of Graphs, vol. 38. American Mathematical Society, Providence (1962)
26. Rozema, L.: Extending the control tower at ShipitSmarter: designing a tool to analyse carrier performance and perform what-if analyses (Master's thesis, University of Twente) (2016)

# A Semantics Extraction Framework for Decision Support in Context-Specific Social Web Networks

Manuela Freire[1,3(✉)], Francisco Antunes[1,2], and João Paulo Costa[1,3]

[1] INESCC – Computer and Systems Engineering Institute of Coimbra, Coimbra, Portugal
[2] Department of Management and Economics, Beira Interior University, Covilhã, Portugal
[3] Faculty of Economics, Coimbra University, Coimbra, Portugal
maria-m-freire@telecom.pt

**Abstract.** We are now part of a networked society, characterized by the intensive use and dependence of information systems that deals with communication and information, to support decision-making. It is thus clear that organizations, in order to interact effectively with their customers, need to manage their communication activities at the level of online channels. Monitoring these communications can contribute to obtain decision support insights, reduce costs, optimize processes, etc. In this work, we semantically studied the discursive exchanges of a Facebook group created by a strawberries' seller, in order to predict, through Social Network Analysis (SNA) and semantic analysis of the *posts*, the quantities to be ordered by customers. The obtained results show that the unstructured data of the Web's speech can be used to support the decision through SNA.

**Keywords:** Social network analysis · Decision support · Web discourse

## 1 Introduction

The use of social web data offers new possibilities for organizations to support daily-based activities. According to Pang et al. [1], "what others think" has always been an important piece of information for most people, during a decision-making process. The social web has made possible, as never before, to directly collect the opinions and experiences (personal and professional) of a wide range of people without any "formal inquiries", thus allowing to change the way we look at the whole decision process, as it will be addressed in Sect. 2. Tollinen et al. [2] argue that social web monitoring, rather than using explicit surveys, provides more objective results on people's intentions. However, according to Murugesan [3], several issues are still open and unsolved, like the management of the social web content (that grows day by day), its heterogeneity and the effectiveness of its extraction, just to mention a few. In addition, as reported by Batagelj et al. [4], the majority of specific social networks (such as enterprise-based) are context limited and ignoring such contexts can impose large constraints on understanding the underlying phenomena or situation.

The main goal of this paper is to present a framework to extract, process, structure and analyze the collected data from context-specific social networks. Such framework, detailed in Sect. 3, incorporates two important facets, namely, human interaction and network structure, by combining human capabilities, Social Network Analysis (SNA) and automatic data

© Springer International Publishing AG 2017
I. Linden et al. (Eds.): ICDSST 2017, LNBIP 282, pp. 133–147, 2017.
DOI: 10.1007/978-3-319-57487-5_10

mining. According to Marmo [5], the combination of SNA and web mining gives an innovative degree of detail in the analysis of social networks that can be useful for decision-making, by providing a better structuring and understanding of the logical sequence of the produced contents of a social web discourse, as expressed by Antunes et al. [6].

However, it is recognized that this type of research is still in its early stages [7], as it requires an intertwined use of linguistics/discourse analysis, natural language processing (NLP), data mining, etc. Most of existing studies turn out to focus on describing network properties, rather than focusing on a deep semantic analysis of the posts and later use of the findings. Although there are more complex studies that deepen the semantic analysis of web discourse contents (see [8–10], for further details), they do not integrate it with the analysis of network properties nor with social actors interaction. The studies that incorporate SNA, as well as the semantic analysis of the posts are scarce and they remain mainly theoretical [11, 12]. In this paper, we pretend to give another step in bridging such gap. To do so, we used the proposed framework to extract, process, structure and analyze the collected data from a social network that was developed around a managerial situation (using Facebook), as described in Sect. 4, where the obtained results, regarding the applicability of the proposed framework, are also discussed. The conclusions of this study are presented in Sect. 5.

## 2   Decision Support in the Social Web Context

One of nowadays most common means to collect data (regarding, for instance, marketing purposes like testing new products acceptance, determining the level of client satisfaction, accessing after-sales quality, etc.) is the use of direct surveys (whether personal or online). Nonetheless, it is largely recognized that this type of approach possesses intrinsic problems, as people do not always reveal their true opinions or intentions, especially in face-to-face situations (whether from politeness, lack of courage, fear of eventual consequences or simply because the interviewer is so nice or beautiful...), which may lead to significant errors in predicting activities such as future sales or poll results.

The Latin expression *in vino veritas* suggests an easy way to solve the earlier problem. Although we do not advocate that we should offer a few drinks to every subject in a statistical sample just to get them "to speak", the idea of collecting people's true opinions seems far more feasible within the context of social webs, than in face-to-face environments, as people can use made-up profiles to express their true ideas, instead of using their "official" profiles (please see Tollinen et al. [2]). In term of the earlier stages of decision-making process (especially at the intelligence and design stages, as defined by Simon [13]), this means getting better information quality, thus enhancing the possibility of better or more reality-tuned decisions, even though direct surveys have their own advantages, such as simplicity, cost and simpler data processing, that social web content analysis does not. There are several limitations, that cannot be disregarded, concerning the treatment, organization and retrieval of the involved information and, in spite of the opinion expressed by Robinson et al. [14] that information processing continues to evolve, the task of collecting and analyzing the content of social web content (or web discourse), commonly known as *posts*, remains quite challenging:

- *Processing the text of posts* – processing the textual data contained in the *posts* requires tools to clean and standardize them (as the ideas contained in *posts* can be diffused along sentences full of characters and punctuation, to emphasize situations), in order to capture the semantic aspects that allow to go beyond the mere identification of keywords, ranking of concepts, etc. [14];
- *Semantics* – usually, *posts* do not share common ontologies, as they are created and changed constantly. The non-existence of standards to express web data semantics hinders the possibilities for integrating applications to analyze them. In addition, the interaction within social networks, and its consequent discursive exchanges, often produces information in an informal and unstructured language, that social tagging, used in folksonomies, fails to address;
- *Post dimension (number of characters)* – in order to accelerate communication people tend to reduce the number of characters that are typed to express an idea [15], which poses greater strain in semantic extraction;
- *Data characteristics* – social networks are much more than a simple set of links and texts. They are interactive and dynamic complex networks with several links that correspond not only to users and "friends" (followers), but also a set of links between *posts*, videos, photos, etc. [14, 16]. In addition, a user can comment *posts* in a successively way (*post → comment*; *comment → comment*) and also share them. This raises

**Table 1.** SNA metrics and their interpretation [17]

| SNA metrics | Interpretation |
|---|---|
| Degree | It is the number of links incident on (pointing to or leaving from) a node and is used to identify nodes that have highest number of connections in the network |
| In-degree | It is the number of links pointing to a node and is used to identify nodes that receive highest number of interactions in the network |
| Out-degree | It is the number of links pointing from a node and is used to identify nodes that send highest number of interactions in the network |
| Closeness | It is the degree of nearness (direct or indirect) between a node and the rest of the nodes in the network. It is the inverse of sum of the shortest distance (also called geodesic distance) between a node and the rest of all in the network |
| Betweenness | It measures the fraction of all shortest paths that pass through a node or in simple terms it quantifies the number of times a node acts as a bridge along the shortest path between two other nodes. Nodes with high betweenness play a crucial role in the information flow and cohesiveness of the network and are considered central and indispensable to the network due to their role in the flow of information in the network |
| Eigenvector | It is a more sophisticated version of degree. It not only depends on the number of incident links but also the quality of those links. This means that having connections with high degree nodes contributes to the eigenvector value of the node in question |
| Page rank | It is a variation of the eigenvector (computed in somehow different way) |

the issues of visualization and graphical analysis. The collected data falls into three categories: structured (users), semi-structured (*posts*) and unstructured (concepts);

- *Data analysis* – the use of mathematical techniques, matrices and graphs, when analyzing social networks, allows to represent and describe the networks in a compact and systematic way, thus accelerating data manipulation [18]. In this case, and according to Jamali et al. [19], SNA can provide a visual (qualitative), as well as a mathematical (quantitative) analysis of human relations. In particular, SNA contemplates several metrics associated with the study of social networks (Table 1 summarizes the SNA metrics used in this work), in spite of the fact that Hanneman et al. [20] point out that there are no right or wrong ways or indicators in approaching social networks.

## 3    Proposed Framework

The proposed framework, depicted in Fig. 1, starts by extracting data from a context-specific social web network. Throughout the process, several networks are generated by an iterative three-stage process, which are recurrent and iterative.

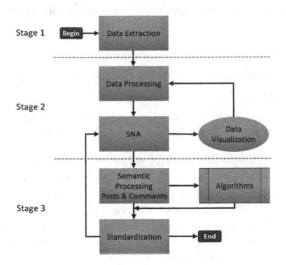

**Fig. 1.** Workflow of the proposed approach (https://meocloud.pt/link/70b41366-2f42-4650-aa7b-ef951839e19c/Fig-01.jpg/)

### 3.1    Data Extraction

The first step is to gather data with all interacting social actors (user → user) and respective *posts* (user → *post*) and data regarding the *posts* and *comments*.

Such data makes it possible to study the social interaction (user → user), as well as to study the semantics of the content of a *post*. To that purpose, a two-mode network is used, as it allows to represent three sets of nodes in the same graph [21], allowing to

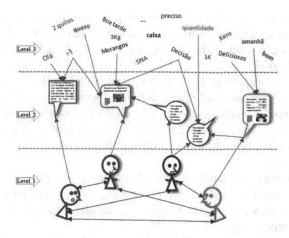

**Fig. 2.** Levels of analysis (https://meocloud.pt/link/0c82fbfa-0726-4d9f-8bf8-3de4ffef39e5/Fig-02.jpg/)

analyze the three levels of interactions between social actors, i.e. the interaction between: users; user and *posts*; *posts* and concepts. The main issue here is to be able to affect a *post* to a user and, in turn, a concept to a *post*, in order to know who said what, as shown in Fig. 2. In this context, concepts are text characteristics that are manually extracted from documents or using preprocessing routines, in order to identify single words, multi-word expressions, whole sentences or syntactic units and are categorized by a specific identifiers [22].

## 3.2 Data Processing and Interpretation

In order to simultaneously analyze the three levels of the network (the interaction between: users; user and *posts*; *posts* and concepts) a transformation of the two-mode network into a one-mode network [23] is required. This method binds the two datasets using common nodes. By doing so, data can be represented by a square matrix, where rows and columns represent the nodes of the two unified datasets. In this matrix a "1" in a given position means that there is a connection between the line node and the column node of that position.

It is then necessary to interpret the data using SNA, namely by using in-degree and out-degree metrics, to eliminate irrelevant data of unconnected nodes, i.e., nodes with in-degree and out-degree equaling zero. These *posts* are usually derived from automatic messages. After this cleaning, it is necessary to perform the junction of the data with all interacting social actors (user → user) and respective *posts* (user → *post* interactions) and data regarding the *posts* and *comments*, in order to create a single network.

## 3.3 Semantics Processing

The first step in this stage is to transform *posts* and *comments*, into networks of words, as a basis to perform the semantic analysis of *posts* using text mining and SNA. The process of data analysis is an iterative recurrent process, where human involvement is essential to systematically analyze the obtained results and to verify if adjustments are necessary. The visualization of the data by means of SNA tools and metrics helps to accelerate this process.

A simple way to transform text into SNA interpretable data is to "split" every discursive exchange (*posts* and *comments*) into individual words. The result of this action is a network where each concept is a node, regardless of its existence in another *post* or *comment*. In doing so, the content of each *post* is summarized through a semantic sub-network, which is constructed by establishing a relationship between the pairs of words contained in each *post*.

In order to identify standardized network concepts, our proposal establishes the creation of a *Cleaning Database* [24] to process the unstructured data from the *posts*. After identifying irrelevant data, the *Cleaning Database* is then configured to discard them in a second processing.

The next step is to detect keywords, since the extraction of keywords is an important technique for identifying the most used and relevant concepts in *posts* [25]. For that purpose, a new network, consisting of the users, received *posts* and concepts contained in them, is then created. In this network each concept constitutes an entity with its own ID. Concepts with the same semantic meaning are then sought (knowing that each concept has a different ID), and the same ID is assigned to all concepts bearing the same meaning, by replicating the links from the keyword to the *post*. We find that the graphical visualization of the network can be very helpful in this step.

A network of keywords is then created, constituted by users, *posts* and concepts contained therein. This network is built with concept's unicity (keywords), allowing to count (using the out-degree metric) the number of times each relevant concept was posted or linked. In addition, variables $k_i$ ($i = 1, \dots, n$) were considered to be the numerical value associated with each of the $i$ concepts. Thus, the global result can be calculated from the product between variable $k_i$ and the out-degree metric for the concept $i$.

## 4   Case-Study: Selling Strawberries

To test and illustrate how the proposed methodology can be used to decision support, a business idea was chosen, namely a "strawberries sale", and implemented with a group of potential customers. A small promotional box of strawberries was delivered at the potential customers' workplaces, with the following attached message: "Directly from the producer to the customer. To place order intentions, please join the Strawberries group created on Facebook and leave yours". Then, a strawberry grower was contacted to deliver those orders at customers' workplaces.

The decision support problem was to estimate the required strawberries quantity to meet the customers' requests (as no pre-established order form was available), solely based on their *posts* on Facebook. We intended to verify if the semantic analysis of the *posts* could be used to effectively recommend the necessary quantities to transport and deliver.

### 4.1  First Stage – Data Extraction

The app *Netvizz* (https://apps.facebook.com/netvizz/), a free tool that provides a simple and quick way to import data through the module "group date", was used to collect the data. The module allowed to extract tabular files of users' activities around the *posts* of the created Facebook group.

The data used in the analysis were gathered over a week. *Gephi* (https://Gephi.org/), an open-source software that allows a simple visualization and manipulation of networks, as well as the calculation of the most important metrics of SNA, was used for data processing. Two distinct sets of data were collected, which allowed the creation of two different networks:

- Network 1 – All social actors with interaction and respective *posts* (user → *post*). This network consisted of 46 nodes and 71 links between them (edges);
- Network 2 – All social actors with interaction (user → user). This network consisted of 13 nodes and 48 links between them (edges).

### 4.2  Second Stage – Data Processing and Interpretation

Facebook contents, captured as pure text, were used to perform a semantic analysis to guarantee the distinction between the discursive textual content, relevant to the analysis, from other forms of language used to communicate. This later type of data, whether character-based (:-), ;(, etc.) or not (images, *smiles*, *emojis*, etc.) required a pre-processing and specific knowledge from other cognition areas (linguistics/discourse analysis, NLP, data mining, etc.), as previously mentioned. The identification and standardization of network concepts was performed using Microsoft Excel VBA (Visual Basic for Applications) based algorithms. The data was then processed using MS-Excel, producing a *Cleaning Database* that was created and imported into *Gephi*, making it possible to generate and visualize the network in a graph.

Initially, the data was entered into *Gephi* as raw data without any processing, to identify irrelevant *posts*. *Gephi* allowed to automatically apply numeric classifications to the network's visual appearance, in terms of the colors and size of entities (users and *posts*) and the modularity class metric was used to identify subgroups [26]. The closeness centrality metrics were used to identify participation differences, as such metrics reveal independence, and shows who communicates with the other network actors, through a minimum number of intermediaries. In order to show the amount of interactions between entities, the in-degree and out-degree metrics were used.

Network 1 had *posts* without links to any other *posts* or users, which were considered irrelevant and eliminated, as they were messages posted by Facebook itself. After deleting those *posts*, data was re-inserted into *Gephi*. Network 3, depicted in Fig. 3(a), derived from Network 1, was formed by all group members, who had at least one connection to a *post*. This network had 36 nodes and 67 edges. Figure 3(b) represents the obtained SNA results. The most active social actor was "ARS", because it had the highest out-degree (10). This meant that he had links between 10 other actors through likes or new *posts*. Actor "CJ" was the element with the lowest out-degree (1), as he only wrote one *post*. Actors with more connections tend to be more powerful because they can directly affect other actors. In Fig. 3(a), the size of the nodes evidences, in a quick and simple way, the importance that the aforementioned actors had within the group.

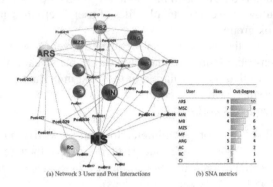

(a) Network 3 User and Post Interactions          (b) SNA metrics

**Fig. 3.** Visualization and results of the user and *post* network, after data mining (https://meocloud.pt/link/a0d36cfd-84ff-4f73-ad19-8ce4b79b1edf/Fig-03.jpg/)

Network 4, resultant from Network 2, consisted of all group elements (users) that had at least one connection to another element, consisting of 13 nodes and 48 edges.

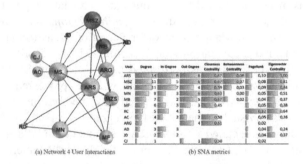

(a) Network 4 User Interactions          (b) SNA metrics

**Fig. 4.** Visualization and results of the user network after data mining (https://meocloud.pt/link/6afc274a-c952-4ca4-ab75-e28374021c64/Fig-04.JPG/)

Figure 4(b) represents the obtained SNA results. The most active social actor was "MSZ" because it had a higher out-degree (6). This means that it has created links between 6 other actors. The actor "ARS" was the element with the highest value (8) in the in-degree metric. This meant that "ARS" *posts* were the most viewed or commented.

"ARS" also had one of the highest values in the metrics closeness centrality, betweenness centrality and eigenvector centrality (0.67, 0.08 and 1 respectively). The fact that s/he had one of the highest values on closeness centrality represents its independence and that it communicated with others through a minimum number of intermediaries. The eigenvector centrality confirmed that "ARS" was one of the most important actors of the network, since this metric aims to measure the importance of a node, according to the importance of its neighbors. Therefore, even if a node is only connected to a few nodes of the network, having a low centrality degree, those neighbors may be relevant and, consequently, the node is also important when it has a high eigenvector centrality.

The "ARS" actor showed a high value in all metrics. This social actor could be designated as a key-player, as it influenced the amount of strawberries that others ordered. This fact is important for web discourse analysis as it indicates that s/he could influence other actors.

In order to obtain a summary of the discursive exchanges, Network 5 was created. This network was a two-mode network transformed into a one-mode network and was constituted by three distinct entities: users, *posts* and concepts. With this transformation, SNA can analyze the three networks to reveal different aspects of the relationships between different entities. To extract the concepts of the *posts*, the *Cleaning Database* and the algorithms created for this purpose were used. The *Cleaning Database* fed the cleaning and standardization algorithms using tables that aid the processing of unstructured data. It was used as a means to: interpret and process the text; extract the semantic network of web speech created by the interaction between social actors; to extract a semantic network to perceive the relation between concepts contained in the text of the discursive exchanges; and to clean and standardize the text, for example by removing spaces and eliminating unnecessary words. As this process can be prone to errors, the process of lexical enrichment of the tables was supported using linguistic tools existing in the NLP literature.

The *Cleaning Database* contained tables of *noise-words*, *"smiles"*, *synonyms* and *punctuation*. It should be noted that the use of an ontology or a *Cleaning Database* must be defined and fed for each decision problem and each decision context.

By visualizing Network 5, Fig. 5, it was possible to identify irrelevant data such as "bom dia" (good morning), "obrigada" (thank you), "olá" (hello), among others. The irrelevant data allowed the reconfiguration of the *Cleaning Database*, by discarding them, in a second processing. The task of identifying this type of data is an iterative recurrent process, where human interaction is essential to systematically analyze the obtained results and to verify if some adjustment is necessary. If so, we must redo everything before moving on to the next step. In this paper it was not possible, due to space constraints, to represent all created and analyzed networks, until the data were completely refined.

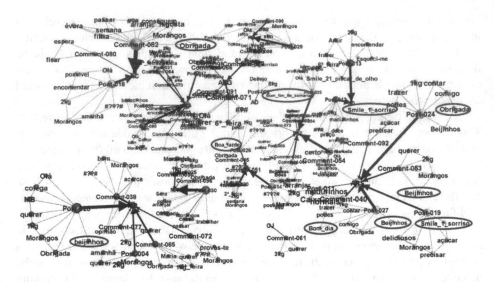

**Fig. 5.** Network 5 Data visualization to support text mining (https://meocloud.pt/link/38b50d55-f9af-4fd7-952d-89b0087a95ad/Fig-5.JPG/)

Finally, Network 6, Fig. 6(a), was created with all the users, *posts* and concepts. Figure 6(b) depicts four examples (A, B, C and D) of semantic subnets, taken from Network 6, summarizing ordering intentions.

### 4.3 Third Stage – Keywords' Network

Individually extracting and analyzing concepts, "keyword", can provide the summary of an opinion expressed in web discourse [1, 25]. Hence, with this goal, Network 8 was created, encompassing users, *posts* and concepts contained therein. Network 8 was built with concept's unicity (keywords) to identify the most intended order quantities (using the out-degree metric) and the customers who intend to place most of the orders.

In order to calculate the quantities intended to be ordered, the out-degree metric was used. We use this metric because we are interested in the node that has the most direct votes, which tells us how many times a keyword has been used. In addition, variables $k_i$ ($i = 1, ..., n$) were considered to be the quantities associated to each of the $i$ concepts. Moreover, the quantities were calculated from the product between variable $k_i$ and the out-degree metric, obtaining the results expressed in Fig. 7(a).

To calculate the quantities ordered by each customer, an attribute that identifies the relationship between "concept → *post* → user" was used. This relationship allowed to calculate the user's out-degree for the concept and, therefore, to identify who wrote the concept and in which *post*. The quantities per customer are the product between the occurrences for each user of the $i$ concept and the variable $k_i$, with the results shown in Fig. 7(b).

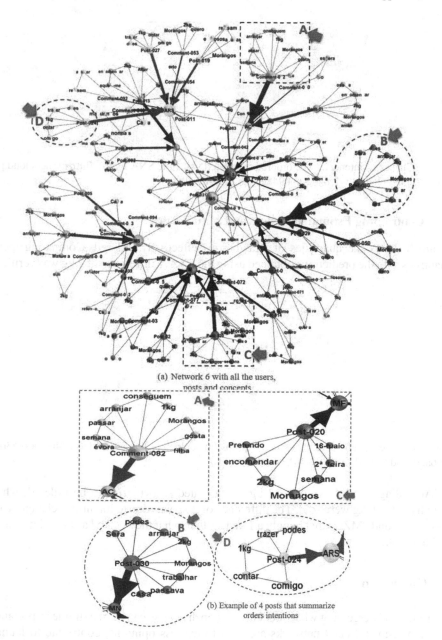

(a) Network 6 with all the users, posts and concepts

(b) Example of 4 posts that summarize orders intentions

**Fig. 6.** Summary of orders intentions (https://meocloud.pt/link/4583ad27-7240-447c-9b20-9e3d020bd0e6/Fig-6a.JPG/; https://meocloud.pt/link/14bdb04e-42ad-47d6-874f-3ad33b324638/Fig-6b.JPG/)

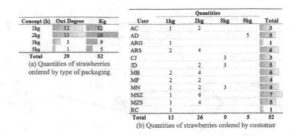

(a) Quantities of strawberries ordered by type of packaging

| Concept (k) | Out-Degree | Kg |
|---|---|---|
| 1kg | 12 | 12 |
| 2kg | 13 | 26 |
| 3kg | 3 | 9 |
| 5kg | 1 | 5 |
| Total | 29 | 52 |

(b) Quantities of strawberries ordered by customer

| User | 1kg | 2kg | 3kg | 5kg | Total |
|---|---|---|---|---|---|
| AC | 1 | 2 | | | 3 |
| AD | | | | 5 | 5 |
| ARG | 1 | | | | 1 |
| ARS | 2 | 4 | | | 6 |
| CJ | | | 3 | | 3 |
| JD | | 2 | 3 | | 5 |
| MB | 2 | 4 | | | 6 |
| MF | 2 | 2 | | | 4 |
| MN | 1 | 2 | 3 | | 6 |
| MSZ | 1 | 6 | | | 7 |
| MZS | 1 | 4 | | | 5 |
| RC | 1 | | | | 1 |
| Total | 12 | 26 | 9 | 5 | 52 |

**Fig. 7.** Quantities intended to be ordered calculated using the SNA metric (https://meocloud.pt/link/03ecbd9c-7edf-4243-84dc-6b0948ecb3b2/Fig-07.JPG/)

### 4.4 Confronting Estimated Orders and Actual Sales

Figure 8 shows the quantities of strawberries effectively ordered. Comparing these quantities with the predictions obtained using the SNA, a difference of 2 kg was verified.

(a) Quantities of strawberries ordered by type of packaging

| Concept (k) | Kg |
|---|---|
| 1Kg | 10 |
| 2Kg | 26 |
| 3Kg | 9 |
| 5Kg | 5 |
| Total | 50 |

(b) Quantities of strawberries ordered by customer

| Cliente | 1Kg | 2Kg | 3Kg | 5Kg | Total |
|---|---|---|---|---|---|
| AC | | 2 | | | 2 |
| AD | | | | 5 | 5 |
| ARG | 1 | | | | 1 |
| ARS | 2 | 4 | | | 6 |
| CJ | | | 3 | | 3 |
| JD | | 2 | 3 | | 5 |
| MB | 2 | 4 | | | 6 |
| MF | 2 | 2 | | | 4 |
| MN | 1 | 2 | 3 | | 6 |
| MSZ | | 6 | | | 6 |
| MZS | 1 | 4 | | | 5 |
| RC | 1 | | | | 1 |
| Total | 10 | 26 | 9 | 5 | 50 |

**Fig. 8.** Orders effectively issued (https://meocloud.pt/link/a8124d5b-e546-4a6b-a3b3-f29ba5049dd7/Fig-08.JPG/)

According to SNA, we got 52 kg of estimated strawberries to be ordered, when actually only 50 kg were sold. The difference of 2 kg was due to intentions change from users "AC" and "MZC" that bought a 1 kg less than what they posted as order intentions on Facebook.

## 5  Conclusions

Capturing and perceiving what circulates in web discourse can be of extreme importance to organizations. Social networks are used to express opinions, contribute to launch ideas, or to react to a situation. Such textual interactions allow to identify *posts* and associated *comments*, in order to segment and identify themes and web discourse semantics, based on the network and *posts'* content. However, because in decision support situations each context is different, it is necessary to adjust data processing features accordingly, and to take into account the language where the web discourse is being produced. Only then, data become relevant and expressive in terms of substance and social context.

This paper presented a framework (methodology, metrics and methods) to extract keywords and summaries of the discursive exchanges of a group. Data were collected, processed and analyzed using SNA tools. We explored a social network, combining two different perspectives: the social interactions between users and the semantic analysis of their discourse. We could conclude that it is possible to extract insights for decision support, namely to predict the behavior of customers.

Regarding social interactions between users, we can conclude that SNA can contribute to the understanding of client behavior, since it allows creating new visions based on real data, resulting from the interaction and discursive exchanges between users. Through SNA metrics it is not only possible to identify the structure of the users' network, but also to "translate" the network into a graphical representation. Such representation makes it possible to identify users acting as leaders (who will make most of the orders) and/or who influence others (product recommendation), as well as to determine the relative strength of the leader through a key concept. Such information can help to devise more adjusted (we could almost say "chirurgical") marketing strategies that target the users that will better propagate its effects on a lesser cost.

Regarding semantic analysis, in order to interconnect concepts to *posts* and *comments*, two approaches were followed. The first one obtained a summary of each discursive exchange, while the second identified the network of keywords. We explored the data with text mining, building semantic networks that summarized the discursive exchanges between users. Text mining allowed to convert *posts* into multiple pieces (as many, as the concepts that exist in a *post*). In addition, the SNA software that incorporated a variety of visualization techniques and algorithms, helped to translate the concepts, extracted from *posts* and *comments*, into something more compact and, therefore, more understandable.

We concluded that not only it was possible to recommend the needed quantities to be transported and delivered to customers, but also we have identified the customers who ordered the most and the leaders of the customers' group.

It should be noted that the construction of datasets obtained from Facebook are inexpensive and easy to extract, so it is easier and faster to obtain reports on the opinions of customers, as compared to conducting market researches or surveys. Increasingly, customers have the ability to express their needs, feelings, desires, and frustrations about a product, service and/or company in real time. Thus, the presented framework has applicability in the prediction of: the number of club fans that will go to a stadium; the number of people attending a demonstration and/or event; the number of people interested in a particular promotional campaign; the number of people who oppose a measure or policy; the acceptance of a new product and/or service, etc. Therefore, we stand that the proposed framework can be useful to extract a set of orientations for decision-making.

Nonetheless, this work presents context-specific limitations, regarding used linguistics and concepts (as it only covers one case study) that might be different in other settings. Additionally, users from other countries use different languages that might have a different structure. For example, the adjective, in English, comes before the noun

("Delicious strawberries"), and, in Portuguese, the adjective comes after the noun ("morangos [strawberries] deliciosos [Delicious]"). In future work it would be interesting to see comparative studies in other context-specific and language to highlight the differences and similarities between semantic extraction frameworks. We intend to further explore semantic data processing algorithms, both in context-specific and context-generic social web networks. Another issue that can be incorporated in the present framework is a "trust metric" for customers. To do so, literature presents several proposals of relationship inference algorithms for assigning a trust score to peers. This confidence between nodes (individuals) of a trusted network makes it possible to calculate a weight granted to the recommendation of the quantity of strawberries to be transported. In future work we intend to incorporate a trust-weight of users (customers) in quantities recommendation.

**Acknowledgments.** This work has been supported by the Portuguese Foundation for Science and Technology (FCT) under project grant UID/MULTI/00308/2013.

# References

1. Pang, B., Lee, L.: Opinion mining and sentiment analysis. Foundations and Trends in Information Retrieval, vol. 2. Now Publishers Inc. (2008)
2. Tollinen, A., Jarvinen, J., Karjaluoto, H.: Opportunities and challenges of social media monitoring in the business to business sector. In: The 4th International Business and Social Science Research Conference, Dubai, UAE (2012)
3. Murugesan, S., Handbook of Research on Web 2.0, 3.0, and X.0 - Technologies, Business, and Social Applications. Advances in E-Business Research Series, pp. 1116. IGI-Global, Hershey (2010)
4. Batagelj, V., et al.: Understanding Large Temporal Networks and Spatial Networks: Exploration, Pattern Searching, Visualization and Network Evolution. Computational and Quantitative Social Science. Wiley, Chichester (2014)
5. Marmo, R., Web mining and social network analysis. In: Zhang, H., Segall, R.S., Cao, M. (eds.) Visual Analytics and Interactive Technologies: Data, Text and Web Mining Applications, pp. 202–211. Information Science Reference, Hershey (2011)
6. Antunes, F., Costa, J.P.: Decision support social network. In: 6th Iberian Conference on Information Systems and Technologies (CISTI), pp. 1–6. Chaves, Portugal (2011)
7. Davenport, T.H.: Big Data at Work: Dispelling the Myths, Uncovering the Opportunities. Harvard Business School Press, USA (2014)
8. Pippal, S., et al.: Data mining in social networking sites: a social media mining approach to generate effective business strategies. Int. J. Innovations Adv. Comput. Sci. 3(2), 22–27 (2014)
9. Makrehchi, M., Kamel, M.S.: A text classification framework with a local feature ranking for learning social networks. In: Seventh IEEE International Conference on Data Mining, Omaha, NE (2007)
10. Gruszecka, M., Pikusa, M.: Using Text Network analysis in corpus studies - a comparative study on the 2010 TU-154 polish air force presidential plane crash newspaper coverage. Int. J. Soc. Sci. Humanity 5(2), 233–236 (2015)
11. Power, D.J., Phillips-Wren, G.: Impact of social media and web 2.0 on decision-making. J. Decis. Syst. 20(3), 249–261 (2012)

12. Herring, S.C.: Discourse in web 2.0: familiar, reconfigured, and emergent. In: Tannen, D., Trester, A.M. (eds.) Discourse 2.0: Language and New Media, pp. 1–25. Georgetown University Press, Washington, DC (2013)

13. Simon, H.A., The New Science of Management Decision. Prentice Hall, Upper Saddle River (1977)

14. Robinson, I., Webber, J., Eifrem, E.: Graph Databases. O'Reilly Media Inc., Gravenstein Highway North (2013)

15. Freire, M., Antunes, F., Costa, J.P.: Exploring social network analysis techniques on decision support. In: 2nd European Conference on Social Media (ECSM 2015), Porto (2015)

16. Herring, S.C.: Web content analysis: expanding the paradigm. In: Hunsinger, J., Klastrup, L., Allen, M. (eds.) The International Handbook of Internet Research, pp. 233–249. Springer, New York (2010)

17. Arif, T.: The mathematics of social network analysis: metrics for academic social networks. Int. J. Comput. Appl. Technol. Res. 4(12), 889–893 (2015)

18. Pinheiro, C.A.R.: Social Network Analysis in Telecomunications. Wiley (2011)

19. Jamali, M., Abolhassani, H.: Different aspects of social network analysis. In: IEEE/WIC/ACM International Conference on Web Intelligence, WI 2006, Hong Kong, pp. 66–72 (2003)

20. Hanneman, R.A., Riddle, M.: Introduction to Social Network Methods. University of California, Riverside (2005)

21. Ikematsu, K., Murata, T.: A fast method for detecting communities from tripartite networks. In: Jatowt, A., Lim, E.-P., Ding, Y., Miura, A., Tezuka, T., Dias, G., Tanaka, K., Flanagin, A., Dai, B.T. (eds.) SocInfo 2013. LNCS, vol. 8238, pp. 192–205. Springer, Cham (2013). doi:10.1007/978-3-319-03260-3_17

22. Feldman, R., Sanger, J.: The Text Mining Handbook: Advanced Approaches in Analyzing Unstructured Data. Cambridge University Press, New York (2007)

23. Borgatti, S.P.: 2-Mode concepts in social network analysis. In: Meyers, R.A. (ed.) Encyclopedia of Complexity and System Science, pp. 8279–8291. Springer, Larkspur (2009)

24. Provost, F., Fawcett, T.: Data Science for Business: What You Need to Know about Data Mining and Data-Analytic Thinking. Data Science for Business. O'Reilly Media, USA (2013)

25. Aggarwal, C.C.: Social Network Data Analytics. Springer, New York (2011)

26. Brandes, U., Erlebach, T. (eds.): Network Analysis - Methodological Foundations. Springer, Heidelberg (2005)

# A Tool for Energy Management and Cost Assessment of Pumps in Waste Water Treatment Plants

Dario Torregrossa[1](✉), Ulrich Leopold[1], Francesc Hernández-Sancho[2], Joachim Hansen[3], Alex Cornelissen[4], and Georges Schutz[4]

[1] Luxembourg Institute of Science and Technology (LIST),
41 Rue du Brill, 4422 Sanem, Luxembourg
dario.torregrossa@list.lu
[2] Estructura Económica, Universitat de València,
Avda dels Tarongers, s/n, 46022 Valencia, Spain
[3] Université du Luxembourg, 6 rue Richard Coudenhove-Kalergi,
1359 Luxembourg Ville, Luxembourg
[4] RTC4Water s.a.r.l, 9, av. des Hauts-Fournaux,
4362 Esch-Sur-Alzette, Belval, Luxembourg

**Abstract.** Waste Water Treatment Plants (WWTPs) are generally considered energy intensive. Substantial energy saving potentials have been identified by several authors. Pumps consume around 12% of the overall WWTP energy consumption. In this paper we propose a methodology that uses the sensors commonly installed in WWTPs, such as volume and energy sensors, to perform energy benchmarking on pumps. The relationship between energy efficiency and flow rate is used to detect specific problems, and potential solutions are proposed, taking into consideration economical and environmental criteria (cost of externalities in energy production). The methodology integrates energy benchmarking, data-mining, and economical and environmental assessment. In order to make better informed decisions, plant managers can now perform a multivariate analysis within a very short time, using information generally available in WWTPs.

**Keywords:** Waste Water Treatment Plant (WWTPs) · Energy benchmarking · Pump performance analysis · Externalities cost

## 1 Introduction

According to [1], Waste Water Treatment Plants (WWTPs) can be considered energy intensive because the energy consumption of the processes are optimised to guarantee an high effluent quality. Several authors demonstrated that there is a high energy saving potential (up to 25%) in WWTPs [2,3]. Pumps are one the major energy consumers in WWTPs using up to 12% of the total energy consumption [4].

© Springer International Publishing AG 2017
I. Linden et al. (Eds.): ICDSST 2017, LNBIP 282, pp. 148–161, 2017.
DOI: 10.1007/978-3-319-57487-5_11

In [5], different causes of inefficiencies are identified (such as pump over-sizing, improper valve set-up, cavitation, wear, mismatching between fluid properties and pump operational characteristics) in order to produce benefits in term of energy saving, maintenance cost and pump life-time. Designing a proper pump system and maintaining the performance during its working life are not trivial tasks, particularly in the WWTP domain, in which, operational conditions are generally not stable because of daily and seasonal patterns, as well as meteorological phenomena in a short-term period and changes in wastewater parameters in a long-term period [6].

In [7], the authors demonstrated the benefits associated to pump parameter monitoring (such as energy consumption, temperature and vibration) for early detection of problems and pump cost minimization (such as energy and maintenance costs).

Even if fundamental, problem detection is not sufficient to identify an optimal solution, because the decision making process should also take into consideration legal restrictions, process requirements and economical and environmental evaluations. In addition, we experienced that although Supervisory Control and Data Acquisition (SCADA) systems [8] often provide access to high time-resolution data, the use of these datasets is often very limited because plant managers generally perform their analysis on a monthly or even a yearly basis. This approach is sub-optimal because the use of aggregated data reduces the quality of the information and because the inefficiencies occurring in the time gap between two plant assessments can produce relevant costs. In this work we extend the performance analysis by including economical and environmental benefits and by increasing the temporal resolution from a monthly to a daily data analysis.

We rely on a fully automated Energy Online System (EOS) described in [9] which is able to gather and store the data of all sensors of a WWTP to provide it for further analysis.

The aim of this paper is therefore to provide a methodology for the on-line benchmarking of pump energy performance, the economical analysis of potential solutions and the analysis of operation scenarios in order to support the plant manager with decision making on these issues.

## 2 Methodology

EOS [9] gathers and stores: (1) data concerning operational parameters, such as volumes and pollution loads; (2) device parameters, such as temperature; (3) current and even rotational speeds of motors. Using these data EOS has algorithms implemented to perform on-line benchmarking of WWTPs.

The proposed methodology combines the use of remote sensors, database management, daily benchmarking and economical analysis of potential actions in order to analyse the pump performances and to assist decision makers.

In the framework of the INNERS and the EdWARDs projects [9,11], data has been collecting daily from WWTPs and stored in the EOS database, in order to calculate Key Performance Indicators (KPIs) for energy assessment and for the evaluation of plant performance. The initial version of EOS enabled the energy benchmarking of the global energy performance of the plants taking into consideration the energy consumption and the operational conditions. In this paper, we propose to further extend EOS for the technical, economical and environmental evaluation of a specific energy consumer: the pump system. First, the observed energy consumption is compared with the theoretical energy consumption calculated at a medium time-resolution (from 5 min to 1 day) by taking into account operational conditions, such as flow and lift. The time-resolution has been defined as 'medium' because, in our experience, the typical time resolution of energy assessment in WWTPs (weeks, months) can be considered low, while the potential energy optimization of highly dynamic devices like pumps with dynamics up to milliseconds can be considered to have a 'high' time resolution. A higher time-resolution can increase the quality of the results because loss of information in aggregation can produce misleading results. For example, the value of the average water flow in a month doesn't take into consideration shorter period flow fluctuations which greatly affect pump efficiency (please see Sect. 3.2). Ideally the performance analysis should be instantaneous, in reality the time-resolution of the sensors is the main limiting factor. This methodology has been tested on 5 min resolution with simulated data. The real plants connected to EOS database currently produce information with a 2 h time resolution.

The difference between the observed and a theoretical energy consumption is subsequently converted in monetary values using the country-specific cost [€/kWh] of energy retrieved from [10]. This value, corresponding to the potential benefit of efficiency action (i.e. the cost of inefficiency) does not take into consideration the environmental and the social impacts that we consider externalities. In order to fill this gap, we use the country-specific cost of externalities for kWh of consumed energy [€/kWh] that can be added to the national energy cost. The result is a new economic value for the energy efficiency improvements that takes into consideration environmental and social benefits. Next, the potential economical saving is compared with the economical cost of alternatives by calculating the threshold cost that a decision maker should accept to pay for a solution (assuming for example a desired pay-back time <2 years). The last step of the methodology consists in testing some hypothesis using the records in the database. In this paper we analyse the scenario 'pump over sizing' for centrifugal fixed speed pumps because this is a relatively common problem (the over-sizing affects 75% of all pumps [5]). Following this approach, the plant managers have a more detailed information set to be combined with their personal knowledge and experience. In fact, the plant managers can evaluate separately the pump system efficiency, the potential economical benefit and environmental impact associated to the energy efficiency actions. Then they have to integrate this assessment with operational, economical and management information that the system does not treat. For example, if an inefficiency is detected, the plant managers should include in the decision process also the previous pump check report, the manpower cost, the age of the pump and the standard of efficiency required by their companies. The concept of the methodology, illustrated in Fig. 1, explains how the different sources of information are integrated.

This approach is highly reproducible because the system requires only a measurement of the energy consumption of a pump and the volume of treated wastewater as well as some static values (pump installed power, cost of the pump, cost of the energy, the lift and the cost structure).

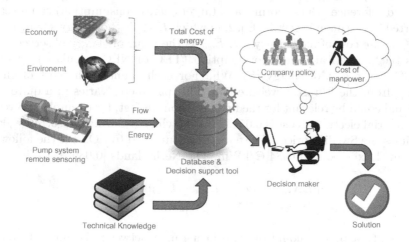

**Fig. 1.** Scheme of the methodology. The on-line sensors provide the information about pump energy consumption and water flow. This information set is combined with economical and environmental information and analyzed with technical knowledge. The tool supports the decision maker in identifying the operational condition of the pump systems. In order to take the final decision, the decision maker integrates this analysis with other plant specific information and with his own experience.

## 3  Pump Assessment

WWTPs are generally equipped with multiple and different pump systems, for example to move the wastewater from the inlet to the primary treatment or for internal recycling and for sludge treatment. The proposed methodology is tested on water pumps but integrating more elements, such as the sludge characteristic, it can be applied also to sludge pumps.

The energy consumption of a pump is generally modelled using Eq. 1. In this equation, Mass is the mass of wastewater [kg], $g$ is the gravity acceleration [9.8 m/s$^2$], H is the lift in meter and $2.78 * 10^{-7}$ is the coefficient to convert joule in kWh. The coefficient $\eta$ accounts the inefficiencies (for example in the conversion of electric energy to rotation energy of the motor, in the pipes, or in the transfer of the rotation energy to the water [12]).

$$E_{th} = \frac{Mass * g * H * 2.78 * 10^{-7}}{\eta} [\text{kWh}] \quad (1)$$

According to [13], $\eta = 0.8$ for high efficient pumps, $\eta = 0.32$ for a normal efficiency while $\eta$ tends to 0.05 for high inefficient pumps. For example, an high efficient pump needs 0.017 kWh to lift 1000 Kg of water 10 m up, while a pump operating normally would consume 0.042 kWh for the same task. The system compares each value of observed energy consumption with a value for normal energy consumption (NEC, $\eta = 0.32$) and a value for the target energy consumption (TEC, $\eta = 0.8$). The difference between an observed energy consumption value and reference values (normal and target energy consumption) can be easily converted into economic cost by Eq. 2, in which $EN_{sp}$ is the energy saving potential, IC is the cost of inefficiency [€], $E_{obs}$ is the value of energy observed kWh, $E_{th}$ is the theoretical energy consumption (TEC or NEC calculated by Eq. 1) and EC is the cost of energy [€/kWh]. For each country, the cost of energy is taken from the Eurostat website [10]. This parameter varies in a quite large range and could be relevant for the final decision. In fact, for example in Sweden, the industrial electricity cost is 0.059 €/kWh, while in Germany is 0.149 €/kWh (2.5 times the Swedish price). The countries connected to EOS are the following: Germany, Luxembourg [0.084 €/kWh], The Netherlands [0.089 €/kWh].

$$\begin{cases} EN_{sp} = max(0, E_{obs} - E_{th})[\text{kWh}] \\ ICreal = EN_{sp} * EC[\text{Euro}] \end{cases} \tag{2}$$

Normally a pump should operate in a range between the normal and the target energy consumption. If the energy consumption is higher than the NEC, the plant manager should react in order to bring the energy consumption to a normal range.

We classify the actions to improve the pump performance in planned maintenance, extraordinary maintenance and device replacement. The planned (routine) maintenance is generally performed by plant managers at fixed-terms and it is not object of this paper. Into the definition of 'extraordinary maintenance', we include all the actions (such as valves regulation, pump cleaning or pump check-up) not included in the planned maintenance except from the pump replacement. The tool presented in this paper supports the decision makers to analyse the pump performance and decide between no-action (action 0), extraordinary maintenance or pump replacement. In order to do this, the tool analyses the cost and the benefit associated to each alternative.

The tool operates under 2 assumptions:

1. the extraordinary maintenance can increase the pump performance up to the normal level ($\eta = 0.32$).
2. the pump replacement is done with the a high efficiency pump, and consequently the energy consumption will meet the target values ($\eta = 0.80$).

### 3.1 Pump Cost Functions

An important parameter to evaluate the benefit of the pump replacement is the cost of the pump system. The tool gives the opportunity for the decision maker to insert the cost of the pump under analysis. If the detailed cost of the pump is not provided by the plant manager, the tool calculates it by the Eq. 3, retrieved from [14,15], in which the P is the installed power [kWh] and f is a coefficient (equal to 1 for pump update, 2 for pump replacement).

$$Cost = f * 16570 * P^{0.559}[\text{Euro}] \qquad (3)$$

Alternatively, the system can estimate the cost of the pump using the maximum volume by extrapolating the values available in [18]. These values, originally available in British Pound for 2006, were converted in € and updated to 2016 with inflation rate. The results are shown in Table 1.

**Table 1.** Pumps installation costs based on maximum inflow

| Maxium flow ($m^3$/day) | k € |
|---|---|
| 4320 | 330 |
| 10800 | 469 |
| 17280 | 694 |
| 864000 | 9823 |

## 3.2   Scenario Analysis

In order to support the decision maker, we want to use the EOS database to detect specific problems (please see Sect. 5), saving time and producing additional information for the decision making process. We propose a methodology to check if centrifugal Fixed Speed Pumps (FSPs) are oversized. According to [16], the hydraulic efficiency $\eta_h$ can be estimated by the Eq. 4, in which q* is the ratio between the current flow and the flow at the best efficiency point (BEP). When $q* > 1$ the pump is working in overload and when $q* < 1$ the pump is working in part load [16]. If a pump normally operates out of an optimal value of q*, the efficiency drops down.

$$\frac{\eta}{\eta_{max}} = 1 - 0.6(q*-0.9)^2 - 0.25(q*-0.9)^3 \qquad (4)$$

# 4   Simulation with STOAT

In order to test the methodology, we used the software STOAT [17], a dynamic simulation model of WWTPs, to simulate the pump behavior. STOAT was chosen because it lets the end user to insert the pump characteristics, the flow time-series and, differently from other softwares, generates the output with a customizable time resolution (from seconds to years). As first step, we generated a flow timeseries, with an average dry weather flow of 100 $m^3$/h and a wet weather flow that can increase up to 500 $m^3$/h. The Fig. 2 reports the shape of the water flow distribution. We have simulated two pump systems. In the first simulation, a single pump is able to manage the maximum flow rate. In this case, the system is expected to operate for most of the time in the part load region. In the second simulation, we combined more pumps activated by a controller according to operational flow condition.

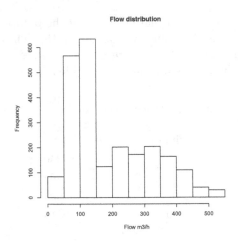

**Fig. 2.** Bimodal distribution reflecting the dry weather flow (left part of histogram) and wet weather flow (right part of histogram). Number of observations = 1734, frequency = 1 observation each 5 min.

### 4.1   Economical Analysis

The economical analysis is developed in different steps: (I) benchmarking, (II) calculation of economical impact of externalities, (III) identification of threshold costs for extraordinary maintenance and pump replacement.

Benchmarking consists in calculating the NEC and the TEC and compare it with the observed data. Figure 3 shows for example the time series of observed data (points in black) compared with the NEC line and with the TEC line. In an efficient system, the observed points should lay between the NEC and the TEC. Next, it is possible to convert the potential energy saving in monetary cost [€] with Eq. 2. This is done for extraordinary maintenance using the NEC and for pump replacement using the TEC.

**Extraordinary Maintenance and Pump Replacement.** The extraordinary maintenance and the pump replacement are evaluated assuming that we want to keep the pay-back time less than 2 years. The maximum cost of an action should not exceed the economical benefit of energy saving in the same time frame. Consequently the threshold can be calculated by up-scaling the cost of inefficiency calculated in the reference period to 2 years using Eq. 5 in which TCP is the threshold cost potential, $avg(IC_{real})$ is the average of the inefficiency

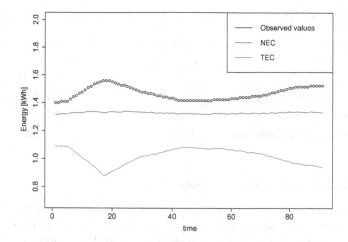

**Fig. 3.** Benchmarking of STOAT pump1. NEC = normal energy consumption, TEC = target energy consumption.

costs calculated for time frame of the period under examination with Eq. 2, $y_n$ is the desired pay-back time in years ($y_n = 2$ in our example), and $N_{day}$ is the time-resolution expressed in day (for example, if the data are available each 5 min $N_{day} = 1/24/60 * 5 = 3.47 * 10^{-3}$, for daily aggregation $N_{day} = 1$). We use the same procedure for pump replacement (assuming $E_{th} = TEC$) and extraordinary maintenance (assuming $E_{th} = NEC$).

$$TCP = \frac{avg(IC_{real})}{N_{day}} * y_n * 365 [\text{Euro}] \tag{5}$$

## 4.2  The Cost of Externalities

The production, the transport and the consumption of energy have also environmental and social impacts. These impacts can be seen as externalities costs or shadow prices [19]. The European Environmental Agency [20] estimated the external costs of Energy Production for each country in Europe taking into account the climate change associated to $CO_2$ emissions, the cost associated to other pollutants (impact on healths, crops...) and the non-environmental costs (such as social costs or costs of the risk associated to nuclear incident). For the countries currently connected to the system the values are the following: 0.020 €/kWh for Germany, 0.018 €/kWh for Luxembourg, 0.001 €/kWh for The Netherlands.

The price of energy is incremented with the cost of externalities and consequently the system takes into consideration the environmental impacts. Table 2 reports the results of these analysis for the first STOAT simulation, and it shows that the use of the external costs increases the cost saving associated to each solution and the decision makers have a bigger profit margin in reducing the inefficiencies.

**Table 2.** STOAT: simulation 1-economical analysis of alternative

| STOAT sim 1 | Average cost saving Euro/day | Threshold cost Euro |
|---|---|---|
| Maintenance | 17.43 | 12,726.04 |
| Maintenance + externalities costs | 21.58 | 15,756.05 |
| Replacement | 31.82 | 23,229.93 |
| Replacement + externalities costs | 39.40 | 28,760.86 |

### 4.3   Data Driven Inefficiency Detection

For each time frame in the database, for fixed speed pumps, the tool plots the $q^*$ and the efficiency as shown in Fig. 4 and assigns each point to different zones. This plot shows the relation between inefficiency and flow conditions. The zones A, B and C, with an absolute $\eta > 0.32$, include the points with a satisfactory performance and the pumps do not require extraordinary maintenance (please note that in the plot we report the relative $\eta$ in order to make a comparison with the theoretical plot calculated with the Eq. 4). The zones D, E and F includes low efficiency points. In the D region, there are the points with low efficiency and $q* < 0.8$ (partload) while in F region there are points with low efficiency and $q* > 1.2$ (overload). The E region corresponds to the points with low efficiency and a normal range for the flow.

**Results of Pump System 1.** For the pump system number 1, Fig. 4 reports the theoretical efficiency depending on the flow and the map of results obtained by simulation. In this case, the system operates mainly in the zone D in which the flow is far for the BEP. In a similar condition, the map in Fig. 4 shows that the pump is oversized. In case of pump system, a similar plot can show a potential error in controller set-up.

**Results of Pump System 2.** In the second simulation, a system of pumps is managed by an automatic controller able to switch on the pumps in order to keep them operating close to their BEP. Figure 4 shows that in this case the distribution of efficiency is not dependent on treated flow. This behaviour can be seen as the validation that the controller operates properly because there are no inefficiencies depending on the flow ratio. Anyway, in this case, since most of the points are in low efficiency zones, a result like the one shown in Fig. 4 suggest to look for other sources of inefficiencies. The Subsect. 5.5 explores more in details the differences between pump analysis and multi-pump system analysis.

## 5   Solution Selection for Fixed Speed Pumps

The study of potential solutions takes into account both the information derived by the scenario analysis (Sect. 3.2) and the economical criteria. The first step

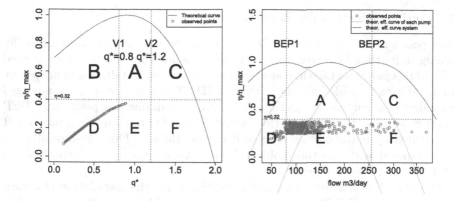

**Fig. 4. subplot left** - Map of the 1st simulation of the STOAT pump. The blue line represent the theoretical behaviour of the pump efficiency. The plot is divided into 6 zones. The horizontal dotted line represent the threshold efficiency $\eta = 0.32$. V1 correspond to $q^* = 0.8$ and V2 line correspond to $q^* = 1.2$. Zone A: flow close to BEP, high efficiency. Zone B: low flow, high efficiency. Zone C: high flow, high efficiency. Zone D: low flow, low efficiency. Zone E: flow close to BEP, low efficiency. Zone F: high flow, ow efficiency. **subplot right** - Map of the 2nd simulation of the STOAT pump system. The blue line represent the theoretical behaviour of the pump efficiency. The plot is divided into 6 zones. The horizontal dotted line represent the threshold efficiency $\eta = 0.32$. The BEP1 corresponds to the BEP of the low-flow pump, and the BEP2 corresponds to the BEP of the high-flow pump. Zone A: flow close to BEFR, high efficiency. Zone B: low flow, high efficiency. Zone C: high flow, high efficiency. Zone D: low flow, low efficiency. Zone E: flow close to BEFR, low efficiency. Zone F: high flow, low efficiency (Color figure online)

consists in identifying the potential problem looking at the maps (like in Fig. 4). Each zone corresponds to an operational condition of the pump system. If the cloud of points belongs to one zone, the pump performance is direct. The points could be also distributed in 2 or 3 zones. In this case, the operator can use these maps with subsamples of the original dataset, for example based on season, in order to investigate other relevant factors for pump performance.

## 5.1   Action Selection for Zones A, B, C

In case the points are highly concentrated in the regions A, B or C the pumps are generally working with absolute $\eta > 0.32$, the Eq. 5 will produce a really low threshold cost potential (TCP) and consequently the tool will propose the decision maker to select the option 0 (no action). Note that the observed points should be under the theoretical curve. A efficient value higher than the one calculated with the Eq. 4 could hide a data problem (for example out-of-ranges, error in aggregation, wrong measurement units...).

## 5.2  Action Selection for Zone D

If the points are concentrated in D zone, it means that the pump is generally working in partflow. Different potential solutions for this problem are reported in [5]. In this paper we take in consideration 2 solutions: installing a 'pony pump' and installing a variable frequency driver (VFD). The pony pump is a pump sized for smaller flow. For example in the first STOAT simulation, the pony pump BEP should be approx. $100 \, \text{m}^3/\text{h}$. In this case, it is possible to estimate a pump cost of 184000 € by extrapolating a value from the Table 1. This value results too high compared with the threshold in Table 2. Consequently the hypothesis to install a pony pump is rejected.

The second alternative taken into consideration is the installation of a variable speed driver (VSD). In this case, we are roughly estimating to install a VSD of 30 kW equal to the peak power demand of the system. An inverter of this power factor costs around 7000 €. This cost is less than the threshold cost reported in Table 2, consequently this solution can be accepted.

## 5.3  Action Selection for Zone E

In the case in which the inefficient points are into the E zone, the inefficiency does not depends on the flow. Other solution should be taken into consideration such as the valve set-up, the pump replacement or the replacement of some other parts of the pump. In this case, it is also strongly suggested to refer to other parameters such as the level of vibrations, the operation temperature or the net positive suction head requirements. In every case, the costs should not exceed the TCP values.

## 5.4  Action Selection for Zone F

With a relevant number of points in the F zone the pumps are under sized. In this case, it is necessary to increase the power of the pump system and the option 0 (no action) is not taken in consideration because the process performance has the priority on economic evaluation.

## 5.5  Differences Between Pump Map and Multi-pump System Map

The same approach, with small changes, can be used to test a multi-pump system. When the tool analyses a set of pumps ruled by a controller, the resulting plot appears like in Fig. 4. In this plot, the theoretical curve is not derived by Eq. 4. A good controller set-up should keep the system efficiency always close to the maximum by using the pumps under their maximum efficiency operational condition. In a multi-pump system, it is not appropriate to define a BEP because this is a pump parameter and because the efficiency should be high in for a flow range (that we call Best Efficient Flow Range, BEFR), ideally corresponding to the operational flow range. Consequently in the x-axis, we report the flow and not a $q^*$.

The BEFR strictly depends on the system configuration (pumps characteristic + controller) and it should be defined case by case. In the second simulation (Fig. 4), in which the controller selects between pumps that not operate at the same time, we identified the BEFR, in the region between the BEP with the pump that operates with low flow (BEP1) and the BEP of the pump that operates with high flow (BEP2). In this example, the efficiency is not depending on the flow (consequently the controller set-up seems to be correct) and the inefficiencies of the points below the threshold should be investigated in different domain (cavitation, wear, valves set-up).

## 6  Discussion

The proposed methodology and tool offer support to decision makers in pump performance optimization with an automatic medium resolution analysis that takes into account the energy consumption, the cost of solutions, the price of the energy and the environmental impacts. The approach is currently able to detect over-sizing for fixed speed centrifugal pumps, an inefficient controller set-up for multi-pump systems and the effects of not-flow-related inefficiencies. A decision maker should be aware that the method works under two important assumptions: the fluctuations in suction head are negligible (i.e. the lift can be considered constant) and the density is considered constant. The methodology proposed in this paper is used to test volume-related issues. The same approach can be used to evaluate the effect of other parameters on the pump performance, for example the water characteristic and/or the temperature.

The method is also able to estimate missing values to perform the analysis, for example by pump cost equation or using the average country value for energy. A potential source of uncertainty is the cost of externalities. The environmental cost of energy consumption greatly depends on the energy source. This analysis can therefore be improved by accounting the energy produced by biogas in the plants differently than the average national mix.

The cost of energy plays a relevant role in the outcome of the final decision because it directly impacts the payback time. The plant managers with low energy cost are less motivated to instigate efficiency actions because the benefit margin is low. For example, for two identical plants in Germany and the Netherlands, the threshold cost will be higher in Germany by a factor of 1.77 (equal to the ratio between energy prices in both countries). This ratio is 1.88 if we also take into account the external cost of energy.

This approach seems more suitable to deal with the pump energy performance than others approaches. In particular the assessment and the classification of operational conditions (Fig. 4) has to be done with knowledge based algorithms because the clusters generated with self-learning methodology [21] could be difficult to analyze in the perspective of decision making. Moreover, when compared with other knowledge-based algorithms (such as decision tree [21]), the graphical approach for the over-size detection seems to be more performing, because it is more intuitive and let to better understand the points close to the edge between 2 regions. Moreover, it is important to remark upon the practical benefit of this methodology because it is able to provide results by using information generally already available in WWTPs.

# 7   Conclusion

The methodology, tested in a simulated environment, promises to provide a medium-resolution analysis based on economical, operational and environmental aspects. The authors are currently working on a more complex model to also take into consideration more details, such as the fluctuation of the suction head and the density variation of wastewater.

In addition, the authors are completing the evaluation of the cost of externalities associated to the installation of a new pump (decommissioning of the old one, the environmental impacts of raw material, transportation costs or benefit for the local economy). A more detailed model has to additionally take into consideration the depreciation of the devices and the additional maintenance costs due to inefficiencies. Moreover, new strategies are under development to investigate different difficulties associated with pumps (motor efficiency, vibration, cavitation and wear). In parallel, the model is under test using actual physical pumps connected to the EOS.

The WWTPs are complex systems and they can be easily equipped with dozen of energy consumer devices (such as pumps and blowers). Consequently, this methodology should be integrated into a global strategy for energy management in WWTPs that takes into account the energy consumption of other devices in order to support the end users in real-time monitoring and managing the full plant. In the next step, the scenario analysis will completed with a list of case-base actions classified according to their specific cost, the response time and the environmental impact. Consequently, the plant managers will be able to understand what is happening in their plants and visualize a list of solutions with associated advantages and disadvantages. The final decision will be taken using the additional information owned by the plant managers and the company policy to prioritize the potential solution according with given criteria (budget, response time and environmental impacts).

**Acknowledgements.** The authors gratefully acknowledge the financial support of the National Research Fund (FNR) in Luxembourg (grant number 7871388-EdWARDS) and the Luxembourg Institute of Science and Technology (LIST).

# References

1. Gude, V.G.: Energy and water autarky of wastewater treatment and power generation systems. Renew. Sustain. Energy Rev. **45**, 52–68 (2015)
2. Castellet, L., Molinos-Senante, M.: Efficiency assessment of WWTPs: a data envelopment analysis approach integrating technical, economic, and environmental issues. J. Environ. Manag. **167**, 160–166 (2016)

3. Becker, M., Hansen, J.: Is the energy-independency already state-of-art at NW-European wastewater treatment plants? In: Proceedings of the IWA conference Asset Management for Enhancing Energy Efficiency In Water And Wastewater Systems, pp. 19–26 (2013)

4. Shi, C.Y.: Mass Flow and Energy Efficiency of Municipal Wastewater Treatment Plants. IWA Publishing, London (2011)

5. Department of Environment (UK): Improving pumping system performance: a sourcebook for industry. Technical report (2006)

6. Water Environment Federation: Operation of Municipal Wastewater Treatment Plants: MoP No. 11, 6th edn. WEF Press, New York (2008). ISBN: 9780071543675. https://www.accessengineeringlibrary.com/browse/operation-of-municipal-wastewater-treatment-plants-mop-no-11-sixth-edition#fullDetails

7. Berge, S.P., Lund, B.F., Ugarelli, R.: Condition monitoring for early failure detection. Frognerparken pumping station as case study. Procedia Eng. **70**(1877), 162–171 (2014)

8. Dieu, B.: Application of the SCADA system in wastewater treatment plants. ISA Trans. **40**(3), 267–281 (2001)

9. Torregrossa, D., Schutz, G., Cornelissen, A., Hernandez-Sancho, F., Hansen, J.: Energy saving in WWTP: daily benchmarking under uncertainty and data availability limitations. Environ. Res. **148**, 330–337 (2016)

10. Eurostat: Energy price statistics (2016). http://ec.europa.eu/eurostat/statistics-explained/index.php/Energy_price_statistics

11. INNERS: Innovative energy recovery strategies in the urban water cycle. Technical report (2015). http://inners.eu/

12. Marchi, A., Simpson, A.R., Ertugrul, N.: Assessing variable speed pump efficiency in water distribution systems. Drink. Water Eng. Sci. **5**(1), 15–21 (2012)

13. Spellman, F.R.: Handbook of Water and Wastewater Treatment Plant Operations. Lewis Publishers, Boca Raton (2013)

14. OECS: The Rural cost functions for water supply and sanitation - Appendix 3: Documentation of expenditure functions - wastewater wastewater. Technical report (2006)

15. COWI, OECD EAP Task Force: Documentation of expenditure functions - wastewater wastewater. Technical report (2004)

16. Gülich, J.F.: Centrifugal Pumps (1998)

17. WRc: STOAT Ver 5.0 (2016). http://www.wrcplc.co.uk/ps-stoat

18. Environment Agency: Cost estimation for control assets summary of evidence report SC080039/R5. Technical report (2015)

19. Fare, R., Grosskopf, S., Lovell, C., Yaisawarng, S.: Derivation of shadow prices for undesirbale outputs: a distance function approach. Rev. Econ. Stat. **75**(2), 374–389 (1993)

20. European Environment Agency: EN35 External costs of electricity production (2008). http://www.eea.europa.eu/data-and-maps/figures/external-costs-of-electricity-production-2

21. Hastie, T., Tibshirani, R., Friedman, J.: The elements of statistical learning. Elements **1**, 337–387 (2009)

# Multi-Criteria Decision Making

# Implementation of an Extended Fuzzy VIKOR Method Based on Triangular and Trapezoidal Fuzzy Linguistic Variables and Alternative Deffuzification Techniques

Nikolaos Ploskas[1]([✉]), Jason Papathanasiou[2], and Georgios Tsaples[2]

[1] Carnegie Mellon University, 5000 Forbes Avenue, Pittsburgh, PA 15213, USA
nploskas@andrew.cmu.edu
[2] University of Macedonia, 156 Egnatia Street, 54006 Thessaloniki, Greece

**Abstract.** Many Multi-Criteria Decision Making (MCDM) problems contain information about the criteria and/or the alternatives that is either unquantifiable or incomplete. Fuzzy set theory has been successfully combined with MCDM methods to deal with imprecision. The fuzzy VIKOR method has been successfully applied in such problems. There are many extensions of this method; some of them utilize triangular fuzzy numbers, while others use trapezoidal fuzzy numbers. In addition, there are many defuzzification techniques available that are used in different variants. The use of each one of these techniques can have a substantial impact on the output of the fuzzy VIKOR method. Hence, we extend the fuzzy VIKOR method in order to allow the use of several defuzzification techniques. In addition, we allow the use of both triangular and trapezoidal fuzzy numbers. In this paper, we also present the implementation of a web-based decision support system that incorporates the fuzzy VIKOR method. Finally, an application of the fuzzy VIKOR method on a facility location problem is presented to highlight the key features of the implemented system.

**Keywords:** Multiple attribute decision making · VIKOR · Fuzzy · Decision support system · Defuzzification

## 1 Introduction

Multi-Criteria Decision Making (MCDM) is a branch of operations research that can be applied for making complex decisions when many criteria are involved. MCDM methods are separated into Multi-Objective Decision Making (MODM) and Multi-Attribute Decision Making (MADM) [15] based on the determination of the alternatives. In MODM, alternatives are not predetermined but instead a set of objective functions is optimized subject to a set of constraints. In MADM, alternatives are predetermined and a limited number of alternatives is to be evaluated against a set of attributes. One of the most widely-used MADM methods is the VIKOR method [7]. The VIKOR method is considered to be effective in cases

I. Linden et al. (Eds.): ICDSST 2017, LNBIP 282, pp. 165–178, 2017.
DOI: 10.1007/978-3-319-57487-5_12

where the decision maker cannot be certain how to express his/her preferences coherently and consistently at the initial stages of the system design [8]. Yu [19] and Zeleny [23] provide the setting theory for compromise solutions. Opricovic and Tzeng [9] state that a compromise solution is a feasible solution, which is closest to the ideal, and compromise means an agreement established by mutual concessions. Hence, the compromise solution can serve as a basis for negotiations. The VIKOR method has been successfully applied on several application areas [18].

There are situations where decision makers have to deal with unquantifiable or incomplete information [22]. Fuzzy set theory can model imprecision in MCDM problems. Variants of the VIKOR method to determine a fuzzy compromise solution in MCDM problems with conflicting and noncommensurable criteria have been developed. In this paper, we utilize a fuzzy extension of the VIKOR method that is based on the methodology proposed by Sanayei et al. [13]; they use trapezoidal fuzzy numbers and in their paper, they focus on supplier selection, but the methodology can easily be applied in a broader scope as well. Variants of the fuzzy VIKOR method with triangular fuzzy numbers have been implemented in the past by Opricovic [10], Rostamzadeh et al. [12], Chen and Wang [2], and Wan et al. [17]. Variants using trapezoidal fuzzy numbers can be found in Shemshadi et al. [14], Ju and Wang [5], and Yucenur and Demirel [20].

In all variants of the fuzzy VIKOR method, a defuzzification technique is necessary to convert fuzzy numbers to crisp values. There are many defuzzification techniques proposed in the literature (for a literature review, see [16]). The use of each one of these techniques can have a substantial impact on the output of the fuzzy VIKOR method. Without trying to propose which method is best, we give the opportunity to decision makers to experiment with different methods and variations and decide which one fits their problem information. Hence, we extend the fuzzy VIKOR method proposed by Sanayei et al. [13] in order to allow the use of several defuzzification techniques. In addition, we allow the use of both triangular and trapezoidal fuzzy numbers. In this paper, we present the implementation of a web-based decision support system that incorporates the fuzzy VIKOR method. Decision makers can easily upload the input data and get illustrative results. Finally, an application of the fuzzy VIKOR method on a facility location problem is presented to highlight the key features of the implemented system.

The remainder of this paper is organized as follows. A background on fuzzy number theory is presented in Sect. 2. In addition, Sect. 2 details the defuzzification techniques that we incorporate in the fuzzy VIKOR method. Section 3 presents the fuzzy VIKOR method. In Sect. 4, the implemented decision support system is presented through a case study on a facility location problem. Finally, the conclusions of this paper are outlined in Sect. 5.

## 2    Background

A fuzzy set is a class with a continuum of membership grades [21]; thus, a fuzzy set $A$ in a referential (universe of discourse) $X$ is characterized by a membership function $A$, which associates with each element $x \in X$ a real number $A(x) \in [0,1]$, having the interpretation $A(x)$ as the membership grade of $x$ in the fuzzy set $A$.

Let's consider now a fuzzy subset of the real line $u: R \rightarrow [0,1]$. $u$ is a fuzzy number [1,3], if it satisfies the following properties:

- $u$ is normal, i.e., $\exists x_0 \in R$ with $u(x_0) = 1$.
- $u$ is fuzzy convex, i.e., $u(tx+(1-t)y) \geq \min\{u(x), u(y)\}, \forall t \in [0,1], x, y \in R$.
- $u$ is upper semi-continuous on R, i.e., $\forall \epsilon > 0, \exists \delta > 0$ such that $u(x) - u(x_0) < \epsilon, |x - x_0| < \delta$.
- $u$ is compactly supported, i.e., $cl\{x \in R; u(x) > 0\}$ is compact, where $cl(A)$ denotes the closure of the set $A$.

The set of elements having the largest degree of membership in $A$ is called the core of $A$:

$$\text{core}(A) = \{x | x \in X \text{ and } \neg (\exists y \in X) (A(y) > A(x))\} \tag{1}$$

The set of all elements that have a nonzero degree of membership in $A$ is called the support of $A$:

$$\text{supp}(A) = \{x | x \in X \text{ and } A(x) > 0)\} \tag{2}$$

One of the most popular shapes of fuzzy numbers is the trapezoidal fuzzy number that can be defined as $A = (\alpha_1, \alpha_2, \alpha_3, \alpha_4)$ with a membership function determined as follows (Fig. 1 (b)):

$$\mu_A(x) = \begin{cases} 0, & x < \alpha_1 \\ \frac{x-\alpha_1}{\alpha_2-\alpha_1}, & \alpha_1 \leq x \leq \alpha_2 \\ 1, & \alpha_2 \leq x \leq \alpha_3 \\ \frac{\alpha_4-x}{\alpha_4}, & \alpha_3 \leq x \leq \alpha_4 \\ 0, & x > \alpha_4 \end{cases} \tag{3}$$

In the case where $\alpha_2 = \alpha_3$, the trapezoidal fuzzy number coincides with a triangular one (Fig. 1(a)).

Given a couple of positive trapezoidal fuzzy numbers $A = (\alpha_1, \alpha_2, \alpha_3, \alpha_4)$ and $B = (b_1, b_2, b_3, b_4)$, the result of the addition and subtraction between trapezoidal fuzzy numbers is also a trapezoidal fuzzy number:

$$\begin{aligned} A(+)B &= (\alpha_1, \alpha_2, \alpha_3, \alpha_4)(+)(b_1, b_2, b_3, b_4) \\ &= (\alpha_1 + b_1, \alpha_2 + b_2, \alpha_3 + b_3, \alpha_4 + b_4) \end{aligned} \tag{4}$$

and

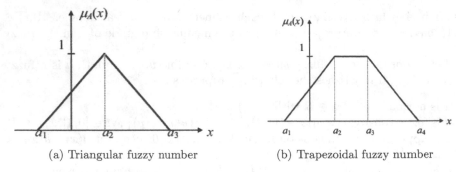

(a) Triangular fuzzy number          (b) Trapezoidal fuzzy number

**Fig. 1.** Triangular (a) and trapezoidal (b) fuzzy numbers, adopted from Lee [6]

$$A(-)B = (\alpha_1, \alpha_2, \alpha_3, \alpha_4)(-)(b_1, b_2, b_3, b_4)$$
$$= (\alpha_1 - b_4, \alpha_2 - b_3, \alpha_3 - b_2, \alpha_4 - b_1) \tag{5}$$

As for multiplication, division, and inverse, the result is not a trapezoidal fuzzy number.

A fuzzy vector is a vector that includes elements having a value between 0 and 1. Bearing this in mind, a fuzzy matrix is a gathering of such vectors. The operations on given fuzzy matrices $A = (\alpha_{ij})$ and $B = (b_{ij})$ are:

– sum

$$A + B = \max [\alpha_{ij}, b_{ij}] \tag{6}$$

– max product

$$A \cdot B = \max_k [\min (\alpha_{ik}, b_{kj})] \tag{7}$$

– scalar product

$$\lambda A \tag{8}$$

where $0 \le \lambda \le 1$.

According to Zadeh [22], a linguistic variable is one whose values are words or sentences in a natural or artificial language. This kind of variables can well be represented by triangular or trapezoidal fuzzy numbers.

All defuzzification techniques can be formulated both in discrete and in continuous form. Without loss of generality and for simplicity, we will use the discrete formulation. The defuzzification techniques that are integrated into the proposed DSS are the following:

– First of maxima (FOM): FOM method selects the smallest element of the core of $A$ as the defuzzification value:

$$FOM(A) = \min (core(A)) \tag{9}$$

- Last of maxima (LOM): LOM method selects the greatest element of the core of $A$ as the defuzzification value:

$$LOM(A) = \max\left(\text{core}(A)\right) \tag{10}$$

- Middle of maxima (MOM): If the core of $A$ contains an odd number of elements, then the middle element of the core is selected such that:

$$\left|\text{core}(A)_{<MOM(A)}\right| = \left|\text{core}(A)_{>MOM(A)}\right| \tag{11}$$

If the core of $A$ contains an even number of elements, then we can select an element as the deffuzification value such that:

$$\left|\text{core}(A)_{<MOM(A)}\right| = \left|\text{core}(A)_{>MOM(A)}\right| \pm 1 \tag{12}$$

- Center of gravity (COG): COG method calculates the center of gravity of the area under the membership function:

$$\frac{\sum_{x_{min}}^{x_{max}} x\mu_A(x)}{\sum_{x_{min}}^{x_{max}} \mu_A(x)} \tag{13}$$

- Mean of maxima (MeOM): MeOM method is a variant of COG method. It computes the mean of all the elements of the core of $A$:

$$MeOM(A) = \frac{\sum_{x \in \text{core}(A)} x}{|\text{core}(A)|} \tag{14}$$

- Basic defuzzification distributions (BADD): BADD method [4] is an extension of the COG method. The defuzzification value is computed as follows:

$$BADD(A) = \frac{\sum_{x_{min}}^{x_{max}} x\mu_A^\gamma(x)}{\sum_{x_{min}}^{x_{max}} \mu_A^\gamma(x)} \tag{15}$$

where $\gamma$ is a free parameter in $[0, \infty)$. The parameter $\gamma$ is used to adjust the method to the following special cases:

$$\begin{cases} BADD(A) = MeOS(A), & \text{if } \gamma = 0 \\ BADD(A) = COG(A), & \text{if } \gamma = 1 \\ BADD(A) = MeOM(A), & \text{if } \gamma \to \infty \end{cases} \tag{16}$$

where $MeOS(A)$ is the mean of support of the core $A$.

## 3    The Fuzzy VIKOR Method

In this paper, we use a fuzzy extension of the VIKOR method that is based on the methodology proposed by Sanayei et al. [13]. This method uses trapezoidal fuzzy numbers and deffuzzifies the fuzzy decision matrix and fuzzy weight of each criterion into crisp values using the COG defuzzification technique (Eq. 13). However, we extend this method by allowing both triangular and trapezoidal fuzzy numbers. In addition, decision makers can also use any of the defuzzification techniques presented in the previous section.

Let us assume that an MADM problem has $m$ alternatives, $A_1, A_2, \cdots, A_m$, and $n$ decision criteria, $C_1, C_2, \cdots, C_n$. Each alternative is evaluated with respect to the $n$ criteria. All the alternative evaluations form a decision matrix $X = (x_{ij})_{m \times n}$. Let $W = (w_1, w_2, \cdots, w_n)$ be the vector of the criteria weights, where $\sum_{j=1}^{n} w_j = 1$.

For simplicity, the steps of the method are presented using trapezoidal fuzzy numbers. The steps of the procedure are:

**Step 1. Identify the objectives of the decision making process and define the problem scope**

The decision goals and the scope of the problem are defined. Then, the objectives of the decision making process are identified.

**Step 2. Arrange the decision making group and define and describe a finite set of relevant attributes**

We form a group of decision makers to identify the criteria and their evaluation scales.

**Step 3. Identify the appropriate linguistic variables**

Choose the appropriate linguistic variables for the importance weights of the criteria and the linguistic ratings for the alternatives with respect to the criteria.

**Step 4. Pull the decision maker opinions to get the aggregated fuzzy weight of criteria and aggregated fuzzy rating of alternatives and construct a fuzzy decision matrix**

Let the fuzzy rating and importance weight of the $k$th decision maker be $\widetilde{x}_{ijk} = (\widetilde{x}_{ijk1}, \widetilde{x}_{ijk2}, \widetilde{x}_{ijk3}, \widetilde{x}_{ijk4})$ and $\widetilde{w}_{ijk} = (\widetilde{w}_{ijk1}, \widetilde{w}_{ijk2}, \widetilde{w}_{ijk3}, \widetilde{w}_{ijk4})$, respectively, where $i = 1, 2, \cdots, m$ and $j = 1, 2, \cdots, n$. Hence, the aggregated fuzzy ratings $(\widetilde{x}_{ij})$ of alternatives with respect to each criterion can be calculated as:

$$\widetilde{x}_{ij} = (x_{ij1}, x_{ij2}, x_{ij3}, x_{ij4}) \tag{17}$$

where:

$$x_{ij1} = \min_k \{x_{ijk1}\}, x_{ij2} = \frac{1}{K} \sum_{k=1}^{K} x_{ijk2},$$
$$x_{ij3} = \frac{1}{K} \sum_{k=1}^{K} x_{ijk3}, x_{ij4} = \max_k \{x_{ijk4}\} \tag{18}$$

The aggregated fuzzy weights $(\widetilde{w}_j)$ of each criterion can be calculated as:

$$\widetilde{w}_j = (w_{j1}, w_{j2}, w_{j3}, w_{j4}) \tag{19}$$

where:

$$w_{j1} = \min_k \{w_{jk1}\}, w_{j2} = \frac{1}{K} \sum_{k=1}^{K} w_{jk2},$$

$$w_{j3} = \frac{1}{K} \sum_{k=1}^{K} w_{jk3}, w_{ij4} = \max_k \{w_{jk4}\}$$

(20)

The problem can be concisely expressed in matrix format as follows:

$$\widetilde{D} = \begin{bmatrix} \widetilde{x}_{11} & \widetilde{x}_{12} & \cdots & \widetilde{x}_{1n} \\ \widetilde{x}_{21} & \widetilde{x}_{22} & \cdots & \widetilde{x}_{2n} \\ \vdots & \vdots & \ddots & \vdots \\ \widetilde{x}_{m1} & \widetilde{x}_{m2} & \cdots & \widetilde{x}_{mn} \end{bmatrix}$$

(21)

and the vector of the criteria weights as:

$$\widetilde{W} = [\widetilde{w}_1, \widetilde{w}_2, \cdots, \widetilde{w}_n]$$

(22)

where $\widetilde{x}_{ij}$ and $\widetilde{w}_j, i = 1, 2, \cdots, m, j = 1, 2, \cdots, n$, are linguistic variables according to Step 3. They can be approximated by the trapezoidal fuzzy numbers $\widetilde{x}_{ij} = (x_{ij1}, x_{ij2}, x_{ij3}, x_{ij4})$ and $\widetilde{w}_j = (w_{j1}, w_{j2}, w_{j3}, w_{j4})$.

**Step 5. Defuzzify the fuzzy decision matrix and fuzzy weight of each criterion into crisp values**

Deffuzzify the fuzzy decision matrix and fuzzy weight of each criterion into crisp values using any of the defuzzification techniques presented in the previous section.

**Step 6. Determine the best and the worst values of all criteria functions**

Determine the best $f_j^*$ and the worst $f_j^-$ values of all criteria functions:

$$f_j^* = \max_i f_{ij}, f_j^- = \min_i f_{ij}, i = 1, 2, \cdots, m, j = 1, 2, \cdots, n$$

(23)

if the $j$th function is to be maximized (benefit) and:

$$f_j^* = \min_i f_{ij}, f_j^- = \max_i f_{ij}, i = 1, 2, \cdots, m, j = 1, 2, \cdots, n$$

(24)

if the $j$th function is to be minimized (cost).

**Step 7. Compute the values $S_i$ and $R_i$**

Compute the values $S_i$ and $R_i$ using the relations:

$$S_i = \sum_{j=1}^{n} w_j \left(f_j^* - f_{ij}\right) / \left(f_j^* - f_j^-\right), i = 1, 2, \cdots, m, j = 1, 2, \cdots, n$$

(25)

$$R_i = \max_j \left[ w_j \left( f_j^* - f_{ij} \right) / \left( f_j^* - f_j^- \right) \right], i = 1, 2, \cdots, m, j = 1, 2, \cdots, n \quad (26)$$

**Step 8. Compute the values $Q_i$**
Compute the values $Q_i$ using the relation:

$$Q_i = v \left( S_i - S^* \right) / \left( S^- - S^* \right) + (1 - v) \left( R_i - R^* \right) / \left( R^- - R^* \right),$$
$$i = 1, 2, \cdots, m \quad (27)$$

where $S^* = \min_i S_i$; $S^- = \max_i S_i$; $R^* = \min_i R_i$; $R^- = \max_i R_i$; and $v$ is introduced as a weight for the strategy of the "maximum group utility", whereas $1 - v$ is the weight of the individual regret.

**Step 9. Rank the alternatives**
Rank the alternatives, sorting by the values $S$, $R$, and $Q$ in ascending order. The results are three ranking lists.

**Step 10. Propose a compromise solution**
Propose as a compromise solution the alternative $[A^{(1)}]$, which is the best ranked by the measure $Q$ (minimum) if the following two conditions are satisfied:

- C1 - Acceptable advantage

$$Q\left( A^{(2)} \right) - Q\left( A^{(1)} \right) \geq DQ \quad (28)$$

  where $A^{(2)}$ is the second ranked alternative by the measure $Q$ and $DQ = 1/(m-1)$.
- C2 - Acceptable stability in decision making: The alternative $A^{(1)}$ must also be the best ranked by $S$ and/or $R$. This compromise solution is stable within a decision making process, which could be the strategy of maximum group utility ($v > 0.5$), or "by consensus" ($v \approx 0.5$), or "with veto" ($v < 0.5$). If one of the conditions is not satisfied, then a set of compromise solutions is proposed, which consists of:
  - Alternatives $A^{(1)}$ and $A^{(2)}$ if only the condition C2 is not satisfied, or
  - Alternatives $A^{(1)}, A^{(2)}, \cdots, A^{(l)}$ if the condition C1 is not satisfied; $A^{(l)}$ is determined by the relation $Q(A^{(l)}) - Q(A^{(1)}) < DQ$ for maximum $l$ (the positions of these alternatives are "in closeness").

## 4    Presentation of the Decision Support System on a Facility Location Problem

The web-based decision support system has been implemented using PHP, MySQL, Ajax, and jQuery. In [11], we presented a DSS that included TOPSIS and VIKOR in a nonfuzzy environment. We extend the DSS to allow decision makers solve group decision-making MADM problems in a fuzzy environment. Next, we present the steps that the decision maker should perform in order to solve a group decision-making MADM problem. More specifically, we consider the facility location (or location - allocation problem). The facility location problem is a well known and extensively studied problem in the operational research discipline. In this case study, a firm is trying to identify the best site out of ten possible choices in order to locate a production facility, taking in the same time into account four criteria: (i) the investment costs, (ii) the employment needs, (iii) the social impact, and (iv) the environmental impact. The first two criteria need to be minimized, while the latter two need to be maximized. The importance weights of the qualitative criteria and the ratings are considered as linguistic variables expressed in positive trapezoidal fuzzy numbers, as shown in Table 1.

**Table 1.** Linguistic variables for the criteria

| Linguistic variables for the importance weight of each criterion | | Linguistic variables for the ratings | |
|---|---|---|---|
| Very low (VL) | (0, 0, 0.1, 0.2) | Very poor (VP) | (0.0, 0.0, 0.1, 0.2) |
| Low (L) | (0.1, 0.2, 0.2, 0.3) | Poor (P) | (0.1, 0.2, 0.2, 0.3) |
| Medium low (ML) | (0.2, 0.3, 0.4, 0.5) | Medium poor (MP) | (0.2, 0.3, 0.4, 0.5) |
| Medium (M) | (0.4, 0.5, 0.5, 0.6) | Fair (F) | (0.4, 0.5, 0.5, 0.6) |
| Medium high (MH) | (0.5, 0.6, 0.7, 0.8) | Medium good (MG) | (0.5, 0.6, 0.7, 0.8) |
| High (H) | (0.7, 0.8, 0.8, 0.9) | Good (G) | (0.7, 0.8, 0.8, 0.9) |
| Very high (VH) | (0.8, 0.9, 1.0, 1.0) | Very good (VG) | (0.8, 0.9, 1.0, 1.0) |

We assume that we have already formed a group of decision makers and one of them acts as the leader of the group. Initially, the leader creates a new MADM problem in the DSS. He/she should enter the following information:

- the name and type (benefit or cost) of each criterion and the name of each alternative (Fig. 2).
- the type of fuzzy numbers (triangular or trapezoidal), the deffuzification technique, and the value of the maximum group utility strategy $(v)$. In addition, the decision maker provides the linguistic variables for the criteria and the alternatives (Fig. 3). The center of gravity defuzzification technique along with trapezoidal fuzzy numbers is used in this case study.

Criteria and alternatives are considered to be evaluated by decision makers that are experts on the field. The evaluations of four decision makers are in Tables 2 and 3. Each decision maker evaluates the criteria and alternatives using a linguistic variable (Fig. 4).

When all decision makers have entered their evaluations, the leader can see the results of the fuzzy VIKOR method. The results are graphically and numerically displayed (Fig. 5). The DSS can also output a thorough report in a pdf file containing the results of the fuzzy VIKOR method. The result is a compromise

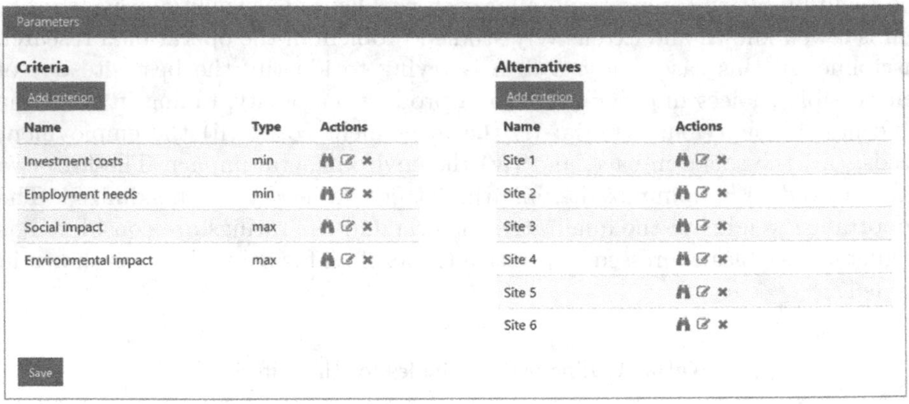

**Fig. 2.** Defining criteria and alternatives

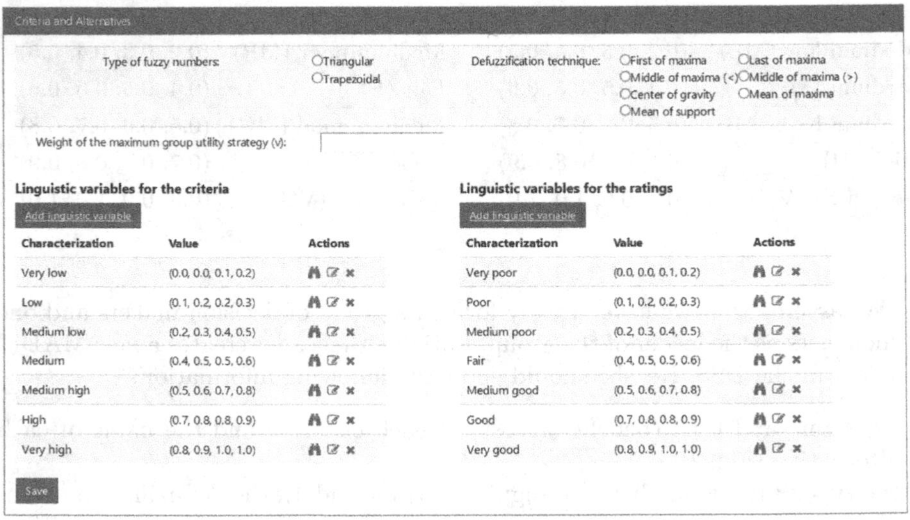

**Fig. 3.** Defining linguistic variables for criteria and alternatives and algorithmic parameters

**Table 2.** The importance weight of the criteria for each decision maker

|  | $D_1$ | $D_2$ | $D_3$ | $D_4$ |
|---|---|---|---|---|
| Investment costs | H | VH | VH | H |
| Employment needs | M | H | VH | H |
| Social impact | M | MH | ML | MH |
| Environmental impact | H | VH | MH | VH |

**Table 3.** The ratings of the ten sites by the four decision makers for the four criteria

| Criteria | Candidate sites | Decision makers | | | | Criteria | Candidate sites | Decision makers | | | |
|---|---|---|---|---|---|---|---|---|---|---|---|
| | | $D_1$ | $D_2$ | $D_3$ | $D_4$ | | | $D_1$ | $D_2$ | $D_3$ | $D_4$ |
| Investment costs | Site 1 | VG | G | MG | MG | Employment needs | Site 1 | F | MG | MG | G |
| | Site 2 | MP | F | F | P | | Site 2 | F | VG | G | G |
| | Site 3 | MG | MP | F | F | | Site 3 | MG | MG | VG | G |
| | Site 4 | MG | VG | VG | MG | | Site 4 | G | G | VG | G |
| | Site 5 | VP | P | G | P | | Site 5 | P | VP | MP | MP |
| | Site 6 | F | G | G | G | | Site 6 | F | MP | MG | MG |
| | Site 7 | P | P | G | P | | Site 7 | VP | P | VP | MP |
| | Site 8 | F | F | F | F | | Site 8 | VG | G | F | G |
| | Site 9 | VG | VG | MG | MG | | Site 9 | P | F | VP | VP |
| | Site 10 | VG | VG | VG | VG | | Site 10 | VG | G | VG | G |
| Social impact | Site 1 | P | P | MP | MP | Environmental impact | Site 1 | G | VG | G | G |
| | Site 2 | MG | VG | G | VG | | Site 2 | MG | F | MP | F |
| | Site 3 | MP | F | F | F | | Site 3 | MP | P | P | P |
| | Site 4 | MG | VG | G | VG | | Site 4 | VP | F | P | F |
| | Site 5 | G | G | VG | G | | Site 5 | G | MG | MG | MG |
| | Site 6 | VG | MG | F | F | | Site 6 | P | MP | F | F |
| | Site 7 | G | VG | VG | VG | | Site 7 | VG | MG | MG | G |
| | Site 8 | MG | MG | G | G | | Site 8 | F | MP | F | G |
| | Site 9 | VG | G | VG | VG | | Site 9 | G | G | VG | VG |
| | Site 10 | VP | F | F | VP | | Site 10 | G | G | VG | VG |

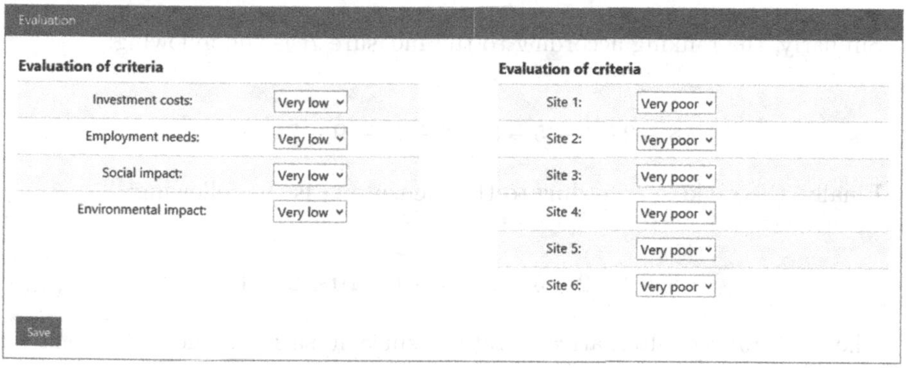

**Fig. 4.** Evaluating criteria and alternatives

solution (if the acceptable advantage condition (C1) and the acceptable stability condition (C2) are met) or a set of compromise solutions.

The final results are shown in Fig. 5. The ranking according to the measure $S$ is the following (first is the most preferred site):

$$8 - 7 - 3 - 5 - 2 - 6 - 1 - 10 - 9 - 4$$

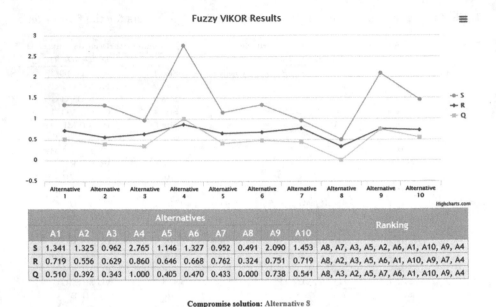

**Fig. 5.** Final results for the facility location problem

Similarly, the ranking according to the measure $R$ is the following:

$$8 - 2 - 3 - 5 - 6 - 1 - 10 - 9 - 7 - 4$$

Finally, the ranking according to the measure $Q$ is the following:

$$8 - 3 - 2 - 5 - 7 - 6 - 1 - 10 - 9 - 4$$

The best ranked alternative is Site 8 since it satisfies the conditions C1 and C2:

- C1: $Q\left(A^{(2)}\right) - Q\left(A^{(1)}\right) \geq DQ \Rightarrow 0.343 - 0 \geq 0.11$ that holds true.
- Site 8 is best ranked by measures $S$, $R$, and $Q$.

## 5    Conclusions

A common problem in many MCDM problems is the existence of unquantifiable or incomplete information about the criteria and/or the alternatives. One way to deal with imprecision is the utilization of fuzzy set theory in the traditional MCDM methods. The fuzzy VIKOR method is gaining popularity in such problems. However, there are many extensions of this method. Some extensions utilize triangular fuzzy numbers, while others use trapezoidal fuzzy numbers. In addition, there are many defuzzification techniques that are used in different variants. In all variants of the fuzzy VIKOR method, a defuzzification technique is necessary to convert fuzzy numbers to crisp values. The use of a defuzzification technique in the fuzzy VIKOR method can have a substantial impact on its output. Hence, it is critical to allow decision makers experiment with different defuzzification techniques.

Without trying to propose which method is best, we give the opportunity to decision makers to experiment with different methods and variations and decide which one fits their problem information. Hence, we extended the fuzzy VIKOR method proposed by Sanayei et al. [13] in order to allow the use of several defuzzification techniques. In addition, we allow the use of both triangular and trapezoidal fuzzy numbers. We also presented the implementation of a web-based decision support system that incorporates the fuzzy VIKOR method. Decision makers can easily upload the input data and get thorough illustrative results. Finally, an application of the fuzzy VIKOR method on a facility location problem was presented to highlight the key features of the implemented system.

# References

1. Bede, B.: Mathematics of Fuzzy Sets and Fuzzy Logic. Springer, Heidelberg (2013)
2. Chen, L.Y., Wang, T.C.: Optimizing partners' choice in IS/IT outsourcing projects: the strategic decision of fuzzy VIKOR. Int. J. Prod. Econ. **120**(1), 233–242 (2009)
3. Diamond, P., Kloeden, P.: Metric topology of fuzzy numbers and fuzzy analysis. In: Dubois, D., Prade, H. (eds.) Fundamentals of Fuzzy Sets, pp. 583–641. Springer, New York (2000)
4. Filev, D.P., Yager, R.R.: A generalized defuzzification method via BAD distributions. Int. J. Intell. Syst. **6**(7), 687–697 (1991)
5. Ju, Y., Wang, A.: Extension of VIKOR method for multi-criteria group decision making problem with linguistic information. Appl. Math. Model. **37**(5), 3112–3125 (2013)
6. Lee, K.H.: First Course on Fuzzy Theory and Applications, vol. 27. Springer Science & Business Media, Heidelberg (2006)
7. Opricovic, S.: Multicriteria optimization of civil engineering systems. Faculty of Civil Engineering, Belgrade **2**(1), 5–21 (1998)
8. Opricovic, S., Tzeng, G.H.: Compromise solution by MCDM methods: a comparative analysis of VIKOR and TOPSIS. Eur. J. Oper. Res. **156**(2), 445–455 (2004)
9. Opricovic, S., Tzeng, G.H.: Extended VIKOR method in comparison with outranking methods. Eur. J. Oper. Res. **178**(2), 514–529 (2007)
10. Opricovic, S.: Fuzzy VIKOR with an application to water resources planning. Expert Syst. Appl. **38**(10), 12983–12990 (2011)
11. Papathanasiou, J., Ploskas, N., Bournaris, T., Manos, B.: A decision support system for multiple criteria alternative ranking using TOPSIS and VIKOR: a case study on social sustainability in agriculture. In: Liu, S., Delibašić, B., Oderanti, F. (eds.) ICDSST 2016. LNBIP, vol. 250, pp. 3–15. Springer, Cham (2016). doi:10.1007/978-3-319-32877-5_1
12. Rostamzadeh, R., Govindan, K., Esmaeili, A., Sabaghi, M.: Application of fuzzy VIKOR for evaluation of green supply chain management practices. Ecol. Ind. **49**, 188–203 (2015)

13. Sanayei, A., Mousavi, S.F., Yazdankhah, A.: Group decision making process for supplier selection with VIKOR under fuzzy environment. Expert Syst. Appl. **37**(1), 24–30 (2010)
14. Shemshadi, A., Shirazi, H., Toreihi, M., Tarokh, M.J.: A fuzzy VIKOR method for supplier selection based on entropy measure for objective weighting. Expert Syst. Appl. **38**(10), 12160–12167 (2011)
15. Tzeng, G.H., Huang, J.J.: Multiple Attribute Decision Making: Methods and Applications. CRC Press, Boca Raton (2011)
16. Van Leekwijck, W., Kerre, E.E.: Defuzzification: criteria and classification. Fuzzy Sets Syst. **108**(2), 159–178 (1999)
17. Wan, S.P., Wang, Q.Y., Dong, J.Y.: The extended VIKOR method for multi-attribute group decision making with triangular intuitionistic fuzzy numbers. Knowl.-Based Syst. **52**, 65–77 (2013)
18. Yazdani, M., Graeml, F.R.: VIKOR and its applications: a state-of-the-art survey. Int. J. Strateg. Decision Sci. (IJSDS) **5**(2), 56–83 (2014)
19. Yu, P.L.: A class of solutions for group decision problems. Manage. Sci. **19**(8), 936–946 (1973)
20. Yucenur, G.N., Demirel, N.Ç.: Group decision making process for insurance company selection problem with extended VIKOR method under fuzzy environment. Expert Syst. Appl. **39**(3), 3702–3707 (2012)
21. Zadeh, L.A.: Fuzzy sets. Inf. Control **8**(3), 338–353 (1965)
22. Zadeh, L.A.: The concept of a linguistic variable and its application to approximate reasoning. Inf. Sci. **8**(3), 199–249 (1975)
23. Zeleny, M.: Multi Criteria Decision Making. McGraw-Hills, New York (1982)

# Integrating System Dynamics with Exploratory MCDA for Robust Decision-Making

Georgios Tsaples[1(✉)], Jason Papathanasiou[1], and Nikolaos Ploskas[2]

[1] University of Macedonia, 156 Egnatia Street, 54006 Thessaloniki, Greece
gtsaples@uom.edu.gr
[2] Carnegie Mellon University, 5000 Forbes Avenue, Pittsburgh, PA 15213, USA

**Abstract.** The aim of this paper is to propose a process to support decision making, in which System Dynamics is combined with Multi Criteria Decision Aid methods to mitigate the limitations of the two methodologies when used alone and find robust policies. The proposed process is based on Exploratory Modeling and Analysis, a framework that allows the use of multiple methods – under different perceptions, detail, and levels of abstraction – in order to address issues of uncertainty and robustness. A case study is used to illustrate how the process can offer deeper insights and act as a valuable decision support system. Finally, it also demonstrates the potential of Exploratory Modeling and Analysis to deal with uncertainties and identify robust policies.

**Keywords:** Exploratory Modeling and Analysis · System Dynamics · Multi Criteria Decision Aid · Decision support system

## 1 Introduction

To support decision-makers cope with the complex nature of decisions, a set of quantitative methods has been used in the literature. These methods provide a formal structuring of the reasons for which a policy is considered a solution to a problem [1]. Two such methods that make use of computational models are System Dynamics (SD) and Multi-Criteria Decision Aid (MCDA).

SD is a methodology that helps to understand the behavior of complex systems over time [2, 3]. Its main elements are feedback loops and time delays that give rise to dynamic complexity [4]. The main goal of SD is to understand how a system's behavior emerges over time and use this understanding to design and test policies in a consequence-free environment [5]. However, SD models do not concern themselves with an explicit evaluation of the policy's performance [5] and whether a policy is preferred over another is often based on the modeler's intuition [6]. Consequently, MCDA can be used to facilitate decision makers to organize the information obtained by the SD models and identify a preferred course of action [5].

MCDA is a branch of Operational Research that aids decision makers to structure a problem, assess alternatives/policies and reach an informed decision [7]. There are many MCDA methods and the choice of one over another depends on the familiarity with the method, the number of criteria, the number of stakeholders involved, the degree of

© Springer International Publishing AG 2017
I. Linden et al. (Eds.): ICDSST 2017, LNBIP 282, pp. 179–192, 2017.
DOI: 10.1007/978-3-319-57487-5_13

compensation [8], the expected outcome [9] and the level of uncertainty of the problem in hand [10].

However, MCDA suffers from several limitations. First, the incorporation of the points of view of diverse stakeholders (supporting different preferences, different criteria etc.) creates substantial problems to an analyst. Second, the large number of MCDA methods makes it difficult to identify and use the most appropriate one. Finally, an MCDA methodology can be considered static, since the evaluation of the policies occurs at specific points in time [11].

In addition, uncertainty is inherent to both decision making and the methods used to facilitate it. Several sources of uncertainty can be identified:

- The difference between the real system "as is" and "as it is perceived to be"
- Sources of uncertainty in the measurement of data
- Uncertainties associated with every method
- The interactions with the decision-makers may cause distortion/additional sources of uncertainty [12, 13].

As a result, the presence of uncertainty on many levels cannot be avoided. Nevertheless, the deployment of these methods is supposed to help decision makers despite uncertainty. Thus, the chosen policy must withstand uncertainty or be robust. In the international literature, several definitions of uncertainty exist, however, in the context of the present paper, the notion of robust conclusion is adopted; where robust means valid in all or most of the versions of the computations (combination of parameters, choices etc.) [14].

One approach to mitigate several of the disadvantages of SD and MCDA when used alone and direct the analysis to a more robust conclusion, is to integrate/combine the two methods [15]. This combination occurs with proposed policies being simulated with the SD model, the identification of criteria and their values from the variables of the SD model and the use of MCDA to provide a structured way of policy assessment [16]. Several attempts focused in the past on the combination of the two methods. In [5], SD and MCDA were combined to create a framework for more efficient performance measurement in organizations. In [6], the authors combined the two methodologies to study the impacts of a construction boom on an urban environment. In [16], the aim was to improve intermodal transportation sustainability. In [17] the purpose is to define appropriate strategies in hypercomplex socio-economic structures, while in [18] the methodologies were used to provide recommendations during a malpractice crisis.

However, there are several limitations in the aforementioned works. First, little effort has been devoted into incorporating different points of view of the system under study. Moreover, the uncertainties associated with each method, how to mitigate their effects and the static nature of MCDA are still a work in progress. Finally, the nature of robustness and its uses in the literature, makes it difficult to identify policies that are valid in almost all versions of the computations. Hence, several sources of uncertainty that could still be mitigated are still present.

The objective of this paper is to combine System Dynamics and MCDA with the purpose of simulating and identifying robust policies, while reducing the effects of several sources of uncertainty. The achievement of the objective will be accomplished by the development of a process, through a computer algorithm.

The rest of the paper is structured as follows: Sect. 2 provides the framework and details of the proposed process. To illustrate the process and assess its potential, a case study is reported in Sect. 3. In Sect. 4 a discussion on the process and future directions of research are presented.

## 2 Methodology

This section provides an overview of the general framework of the proposed process and its specific details.

### 2.1 Exploratory Modeling and Analysis

As already mentioned, the large number of MCDA methods and the difficulties associated with the choice of one, led the research towards the notion of "satisfaction of the decision maker" [8]. This approach led to the use of more than one MCDA methods for validation purposes [19].

However, the use (or the integration) of more than one modeling methods, falls into the category of Exploratory Modeling and Analysis (EMA). EMA is a framework and a methodology that relies on computational power to analyze deep uncertainties [20]. In contrast to conventional predictive modeling, where a single model is used to predict a system's behavior, EMA acknowledges that building such a model is impossible in the context of long term policy analysis. Thus, under the EMA framework, different models of the same system can be built under different levels of detail, abstraction and points of perception. These models can then be explored under computational experiments in an effort to reveal insights of the uncertainties [21]. In the context of SD modeling, the EMA framework has been used in many works e.g. [15, 21]. In the MCDA field a first effort to use EMA was performed in [19].

In the rest of this section, an overview of the proposed process is described.

### 2.2 Exploratory MCDA

The first step in the process is the development and simulation of an SD model that describes the system under study. Next, the policies are designed and simulated in the model and the generated results will provide the data for the next steps. In more detail, the model will provide the set of criteria from which the decision makers can choose. Furthermore, instead of making assumptions on the values of those criteria or conducting interviews with experts or using any kind of methods that are typically used in the MCDA context, the SD model will provide these values in different time periods. These policies will be the same for all decision makers and known at the beginning of the analysis. Furthermore, to avoid the generation of big data, the number of policies is limited.

To take into account different perceptions of a system means to incorporate different points of view of the system "as is perceived to be". Hence, any number of decision makers can be identified, with a different set of criteria and preferences and the common policies will be evaluated separately for each one. That way it is secured that the system and the proposed policies will be examined under different perspectives and approaches on what consists a satisfactory solution.

Regarding the criteria, each one can be an element of the SD model (stock, flow or auxiliary) and its values will be generated by the simulations. Each decision maker can identify different sets of criteria and choose whether they need maximization or minimization.

In addition, more than one MCDA methods will be used for the evaluation of the simulated policies. These methods are: performance targets, SMART and PROMETHEE II. The choice of the particular methods covers the criteria of the classification proposed by [22] and their popularity [23].

For the performance targets, a policy is required to meet a minimum threshold on every criterion. Thus, each decision maker will provide a range of targets/thresholds for each of their respective criteria. Consequently, the policies will be tested for all possible combinations of these thresholds for all the criteria. When all the policies have been tested against all combinations of the thresholds, the process will calculate the number of times each policy failed i.e., at least one of the criteria failed to meet the threshold in each of the different combinations. The policy that failed the largest number of times along with the policies that have a number of failures that falls within a range of a certain percentage of the policy with the largest number of failures, will be excluded from the following steps of the analysis. The testing against the thresholds will be performed for every time step of the SD model; thus, the decision-maker could investigate which policies perform satisfactory on which points in (the simulation) time.

The limit can be chosen arbitrarily by each decision maker and by choosing such a limit, it is ensured that different number of policies might succeed the performance targets. For illustrative purposes in the context of this paper this limit is set to 75% of the policy with the largest number of failures. Furthermore, the limit could be changed to meet the needs of a particular analysis. For example, instead of two groups of policies, there could be three that are separated by the limits of 10%, 25% and the rest. In conclusion, the performance targets offer an opportunity of an initial screening of the simulated policies. Only those that appear robust against the combination of thresholds that each decision maker provides, will continue to the next steps of the analysis.

In SMART, a set of utility functions that are sufficient for most of the cases [24] are:

$$U(x) = a - b^*exp(-c^*x) \tag{1}$$

$$U(x) = a + b^*(c^*x) \tag{2}$$

$$U(x) = a + b^*exp(c^*x) \tag{3}$$

The shape of the Eqs. (1), (2) and (3) is concave, linear and convex respectively, indicating the attitude of the decision maker towards risk (risk averse, neutral and seeking, respectively). In the context of the proposed process, the decision-maker can provide the utility function for each criterion in the form of Eqs. (1), (2) and (3), by providing a range of values for the parameters a, b and c (which are variables that determine the shape of the equations)-assuming that the shape of the equations does not change during the various time steps of the simulation. Moreover, a wide range of parameters can be provided and the process will calculate all possible combinations for every utility function for all criteria.

PROMETHEE II has been used despite the lack of the incomparability notion of PROMETHEE I, because the complete ranking facilitates the aggregation of the results with those obtained with the SMART method. In the context of the proposed process, the decision maker can provide the preference function for each criterion, assuming that it does not change for the analysis. However, the parameters of each preference function can change each time step. The decision maker can provide a range of values for each parameter (per time step) and the process will calculate all possible combinations and will produce all possible rankings.

Finally, regarding the weights, each decision-maker can provide a range of weights (values) for each criterion. The process will calculate all the possible combinations and will take into account only those combinations that lead to $\sum_i^n w_i = 1$, where n is the number of criteria. The notion of weight has different meanings for SMART (level of compensation among the criteria) and PROMETHEE (values of importance). Despite the different meaning, the same weights are used for both the methods in the process. The notion however, is not distorted; the large set of values that is swept during the calculations and the numerous combinations ensure that the preferences of the decision maker are met.

The exploratory MCDA process was developed with the Python programming language. The process is interactive and each time point, the decision maker provides the values that are asked by the program. Figure 1 illustrates an example of the questions asked by the program (ending at the question mark) and the various values that are required by the decision maker during the execution of the process.

```
Weight linspace for Crit4 (start, end, number)=? 0.15,0.75, 3
Give marginal utility function for Crit4=? Risk_neutral(a,b,c)
Factor a linspace values for Crit4 (start, end,number)=? (1, 2, 2)
Factor b linspace values for Crit4 (start, end,number)=? (0.5, 0.7,3)
Factor c linspace values for Crit4 (start, end,number)=? (-0.01, 0.1, 5)
Give preference function for Crit4=? U_form_min(q)
```

**Fig. 1.** Example of the values that are required by the decision-maker during the execution of the process. The numbers are just an example of what a decision-maker could provide, and no actual process took place. For example, for criterion 4, the decision maker has decided to give a minimum weight of 0.15, a maximum of 0.75 and 3 values in between. The process will keep only those values that in addition to the other criteria, satisfy $\sum_i^n w_i = 1$, where n is the number of criteria

Finally, in accordance with the definition provided in the introductory section, the robustness of the simulated policies is studied under the notions of "similarity" or "closeness". Since many rankings will be created (each one for a different combination of the input parameters), two sets of policies will be defined: the first one containing those that have a score higher or equal to the 75% of the policy with the highest score in the specific ranking and the second set will contain the rest of the policies. The limit is chosen arbitrarily for illustration purposes, but it could be modified to address the needs of a specific analysis.

Subsequently, it will be calculated how many times each policy appeared in each set and the policies with the highest number of appearances in the first set will be considered the most robust. Hence, the decision maker could have an overview not only of the consistently "good" or "bad" policies, but also of those that appear to be sensitive under the different combinations of parameters.

To conclude this section, it should be stated that the choices made for the development of the process might not be ideal. For example, a trade-off had to be made: the number of policies that can be simulated in the System Dynamics model is limited; thus the model avoids the generation of excessive amount of data. On the other hand, this is compensated by the capability to include any number of separate decision-makers in the process, ensuring that the system is studies under different perceptions. However, it is a pilot methodology that could be incorporated and transformed into a Decision Support System that could be adapted to the needs of the decision-maker and increase the confidence in the decision-making process. Figure 2 illustrates the flow of the proposed process.

## 3   Case Study

To illustrate the process, a model by Tsaples et al. [25] is used. The model is concerned with urban development and especially the stagnation that is caused by the depletion of the available land and the aging process. The model and the data are generic and experimental (no decision-making process or workshop took place; the numbers provided below are examples that the authors used to illustrate the capabilities of the proposed method). Similar for the exploratory MCDA part of the analysis, which only illustrates the potential of the proposed process. The particular model was chosen because it represents an illustration of a political decision that affects a large number of people, its outcome is uncertain and different stakeholders can be involved in the decision making process and affect its outcome.

Many SD works regarding urban development exist and the developed model is based on the work by [26]. The simulated urban environment is divided into four zones (zone 1 to zone 4). Each zone has a housing sector, a business sector, a population of workers and unemployed and finally, a simplified economic system that is applied to the entire urban region, whose main elements are the three taxes (tax on income, tax on business and land value tax). The population of each zone (workers and unemployed) can move among the zones based on how attractive the destination zone is compared to the origin. Similarly, the population can move in and out of the entire urban environment based on the notion of relative attractiveness. Figure 3 illustrates the main elements of the model and the relationships among them. For example, the population of each zone affects and is affected by the housing availability in the zone; the larger the population in the zone the smaller the housing availability (a relation where an increase/decrease in one variable causes a decrease/increase in another is denoted by the - in the causal

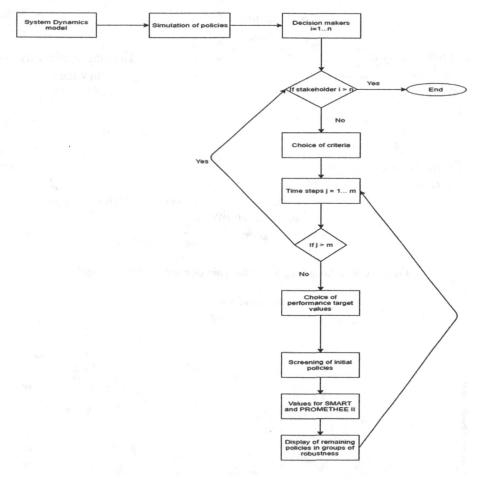

**Fig. 2.** Flowchart of the proposed process

arrow), but as the housing availability falls, the zone becomes unattractive for the population, thus the population falls (that type of relation where an increase/decrease in one variable causes an increase/decrease in another is denoted with the + sign)

The policies that are simulated in the model are different combination of taxes in different points in the simulation time. They are summarized in the Table 1.

The names of the policies are identified in the first column and the rest of the columns show how the values for the three taxes can change during the simulation time. The numbers represent the tax as a percentage on the value of the entity (business, building etc.) on which they are applied. For example, *alt3* is a policy where the land value tax is increased from 0.1 to o.3 in simulation time of 50 (years), while the other two taxes are nullified at the same time.

Some indicative results are shown in the Figs. 4 and 5.

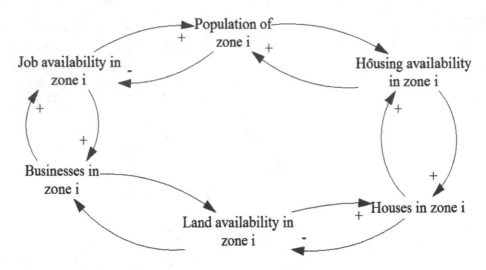

**Fig. 3.** Causal Loop Diagram of the main elements of the SD model

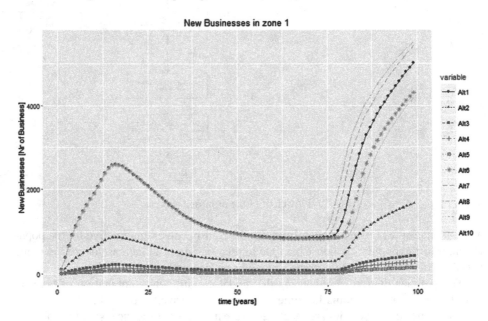

**Fig. 4.** New business under the different policies for zone 1

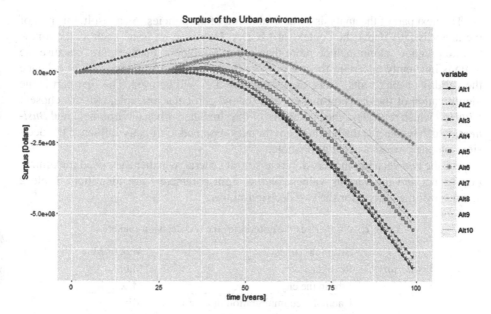

**Fig. 5.** Surplus of the urban environment for the different policies

**Table 1.** Summary of the simulated policies tested in the SD model

|  | Land Value tax | Property tax | Business tax |
|---|---|---|---|
| alt1 | 0.1 | 0.1 | 0.1 |
| alt2 | 0.2 | 0 | 0 |
| alt3 | 0.1 + step(0.2,50) | 0.1 − step(0.1,50) | 0.1 − step(0.1,50) |
| alt4 | 0.1 + step(0.2, 25) | 0.1 − step(0.1, 25) | 0.1 − step(0.1, 25) |
| alt5 | 0.1 + step(0.2, 25) | 0.1 | 0.1 |
| alt6 | 0.1 + step(0.4, 25) | 0.1 + step(0.2, 25) | 0.1 + step(0.2, 25) |
| alt7 | 0.1 + step(0.4, 25) | 0.1 − step(0.1, 25) | 0.1 − step(0.1, 25) |
| alt8 | 0.1 + step(0.2, 10) | 0.1 − step(0.1,10) | 0.1 − step(0.1,10) |
| alt9 | 0.1 + step(0.4 10) | 0.1 − step(0.1, 10) | 0.1 − step(0.1, 10) |
| alt10 | 0.1 + step(0.2,10) +step(0.2, 50) | 0.1 + step(0.2, 10) − step(0.3, 50) | 0.1 + step(0.2, 10) − step(0.3, 50) |

The upper graph illustrates the number of new businesses that are created in a period of 100 years for zone 1 of the simulated urban environment. The different lines in the graph show how the number changes with the different simulated policies. Similarly, the graph on the bottom illustrates the behavior of the city's surplus for the same period under the different policies. Hence, it can be observed that deciding which policy is the most beneficial is not easy based solely on the results of the simulation; a situation which becomes more complex if more than one decision-makers, with different objectives, are considered.

The next part of the analysis is the evaluation of the policies. As a result, this part of the analysis falls under the context of MCDA. However, an issue such as the a new taxation regime should include more than one stakeholders. Furthermore, because the issue is inherently uncertain the use of more than one MCDA methods could enhance the validity of the results. Thus, exploratory MCDA will be used. The first part is the identification of the decision-makers and for the particular example, two are chosen: *CityHall* which represents the authorities of the simulated urban environment and *Business,* which represents the business community-assumed to directly influence the decision-making process and act as a whole.

For the two decision-makers different sets of criteria, whether they need maximization or not and the various values for the input parameters are identified. Table 2 summarizes the criteria for the two decision-makers.

**Table 2.**  summarizes the criteria for the two decision-makers.

|  | Criterion name | Max or Min |
|---|---|---|
| *CityHall* | Revenues | Max |
|  | Jobs in the city | Max |
|  | Unemployed immigration in zone 1 | Min |
|  | Unemployed population in zone 1 | Min |
|  | Slum houses in zone 1 | Min |
| *Business* | New Business value in zone 1 | Max |
|  | Unemployed immigration in zone 1 | Max |
|  | Jobs in the city | Min |
|  | Business tax multiplier | Min |

The two decision-makers have different objectives, which are operationalized by different sets of (conflicting) criteria. Similarly, the input parameters cover a wide range of values for all methods used in the exploratory part of the analysis. Moreover, the times for which the evaluation of the policies is performed is at year 50 and 100 of the simulation time. Finally, the whole process was programmed in python.

The results for the decision-maker *CityHall* are depicted in Fig. 6.

The graph on the left shows the two sets of policies for time 50 of the simulation. Firstly, it can be observed that from the 10 original simulated policies only 5 have passed the first stage of the performance targets. Second, the policies that passed are further divided on Group 1 (large number of high ranking) and Group 2. The color in the columns of each policy, shows the total number the policy appeared in Group 1. Thus, for *CityHall* on time 50 only *alt6* and *alt10* are preferred. However, *alt10* did not pass the performance targets for time 100. Hence, for *CityHall* the policy named *alt6* seems to appear robust under all combinations of parameters (Fig. 6).

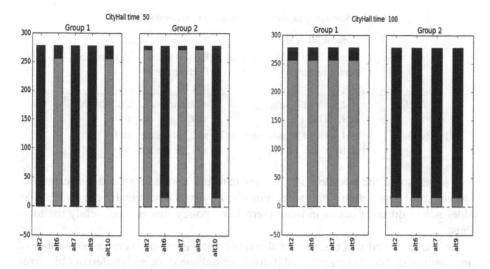

**Fig. 6.** Results for CityHall for time 50 (left) and 100 (right)

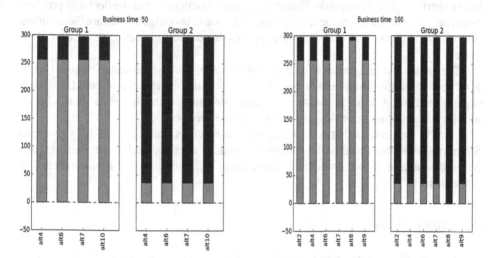

**Fig. 7.** Results for *Business* for time 50 (left) and 100 (right)

For *Business* on the other hand, only 4 simulated policies passed the performance targets at time 50 and all show similar robustness. For time 100 though, *alt10* once again does not pass the performance targets and two new policies are added. From those, the most robust is *alt8,* which did not pass the initial screening at time 50.

The results for the two decision-makers are summarized in the Table 3.

Table 3. Summary of the results for the two decision-makers

|  | Time = 50 years | | Time = 100 years | |
| --- | --- | --- | --- | --- |
|  | Passed Performance targets | More robust | Passed Performance targets | More robust |
| *CItyHall* | alt2, alt6, alt7, alt9, alt10 | alt6 | alt2, alt6, alt7, alt9 | alt2, alt6, alt7, alt9 |
| *Business* | alt4, alt6, alt7, alt10 | alt4, alt6, alt7, alt10 | alt2, alt4 alt6, alt7, alt8, alt9 | alt8 |

It appears that the two decision-makers can find common ground if the process involves some negotiation. Otherwise, it can be observed that even for the same stakeholder across different points in time, there is no policy that is consistently the most robust.

For the proposed process, one of the limitations is on the very large number of combinations that could be generated if different stakeholders, with different objectives and at different times are inserted in the process. That overflow of information could lead to performance downgrade. However, a careful structuring of the decision problem could help decision-makers identify robust policies and investigate what are the common interests and points of friction with other parties that do not share their perception of the system.

The values and preference of each decision-maker had to be inserted manually; this act of collaboration could help to keep preferences and objectives structured and well-organized. Thus, a decision-maker is not only presented with the results, but is forced to logically structure priorities, objectives and preferences [27].

Finally, the process that was presented can serve as the backbone of a Decision Support System that could be adapted to the needs of decision-makers or situations and be used to explore robust policies in hypercomplex, uncertain and multi-stakeholder situations.

## 4    Conclusions

The purpose of this paper was to investigate the integration of System Dynamics with MCDA under the framework of Exploratory Modeling and Analysis and to develop a process to address the disadvantages that exist when these methods are used individually and identify robust policies. A case study showed that the process could offer insights into the identification of robust policies and policies that could be points of friction among different parties in the decision-making process.

The choices that were not made during the development of the process show that there are different approaches to a structured decision-making process, while the choices that were made demonstrate potential avenues of future research and improvement.

Finally, the rise of computational power makes Exploratory Modeling and Analysis more feasible to use; although not without its limitations, it could offer great insights into the decision-making process and the paper is a demonstration of its potential.

# References

1. Ouerdane, W., Maudet, N., Tsoukias, A.: Argumentation theory and decision aiding. In: Ehrgott, M., Figueira, J., Salvatore, G. (eds.) Trends in Multiple Criteria Decision Analysis, pp. 177–208. Springer, US (2010)
2. Forrester, J.W.: Industrial Dynamics. MIT Press, Cambridge (1961)
3. Sterman, J.D.: Business Dynamics: Systems Thinking and Modeling for a Complex World. Irvin/McGraw-Hill, Boston (2000)
4. Sterman, J.D.: Modeling managerial behavior: misperceptions of feedback in a dynamic decision making experiment. Manage. Sci. **35**(3), 321–339 (1989)
5. Santos, V., Belton, V., Howick, S.: Adding value to performance measurement by using system dynamics and multicriteria analysis. Int. J. Oper. Prod. Manag. **22**(11), 1246–1272 (2002)
6. Gardiner, P., Ford, A.: Which policy run is best, and who says so. TIMS Stud. Manag. Sci. **14**, 241–257 (1980)
7. Belton, V., Stewart, T.: Multiple Criteria Decision Analysis: An Integrated Approach. Springer, Heidelberg (2002)
8. Guitouni, A., Martel, J.: Tentative guidelines to help choosing an appropriate MCDA method. Eur. J. Oper. Res. **109**(2), 501–521 (1998)
9. Kunsch, P.L.: How System dynamics education may enhance virtue-based ethics. EURO J. Decision Processes **4**(1–2), 33–52 (2016)
10. Yatsalo, B., Didenko, V., Gritsuyuk, S., Sullivan, T.: Decerns: a framework for multi-criteria decision analysis. Int. J. Comput. Intell. Syst. **8**(3), 467–489 (2016)
11. Polatidis, H., Haralambopoulos, D., Vreeker, R.: Selecting an appropriate multi-criteria decision analysis technique for renewable energy planning. Energy Sources Part B **1**(2), 181–193 (2006)
12. Bouyssou, D.: Modelling inaccurate determination, uncertainty, imprecision using multiple criteria. In: Lockett, A.G., Islei, G. (eds.) Improving Decision Making in Organizations, pp. 78–87. Springer, Heidelberg (1989)
13. Roy, B.: Main sources of inaccurate determination, uncertainty and imprecision in decision models. Math. Comput. Model. **12**(10), 1245–1254 (1989)
14. Roy, B.: A missing link in operational research decision aiding: robustness analysis. Found. Comput. Decision Sci. **23**, 141–160 (1998)
15. Pruyt, E.: Dealing with uncertainties? Combining system dynamics with multiple criteria decision analysis or with exploratory modeling. In: Proceedings of the 25th International Conference of the System Dynamics Society, Boston, USA (2007)
16. Pubule, J., Blumberga, A., Romagnoli, F., Blumberga, D.: Finding an optimal solution for biowaste management in the Baltic States. J. Clean. Prod. **88**, 214–223 (2015)
17. Brans, J., Macharis, C., Kunsch, P., Chevalier, A.: Combining multicriteria decision aid and system dynamics for the control of socio-economic processes: an iterative real-time procedure. Eur. J. Oper. Res. **109**(2), 428–441 (1998)
18. Reagan-Cirincione, P., Schuman, S., Richardson, G., Dorf, S.: Decision modeling: tools for strategic thinking. Interfaces **21**(6), 52–65 (1991)
19. van der Paas, J., Walker, W., Marchau, V., van Wee, G., Agusdinata, D.: Exploratory mcda for handling deep uncertainties: the case of intelligent speed adaptation implementation. J. Multi-Criteria Decision Anal. **17**(1–2), 1–23 (2010)
20. Bankes, S.: Exploratory modeling for policy analysis. Oper. Res. **41**(3), 435–449 (1993)

21. Auping, W., Pruyt, E., de Jong, S., Kwakkel, J.: The geopolitical impact of the shale revolution: Exploring consequences on energy prices and rentier states. Energy Policy **98**, 390–399 (2016)
22. Roy, B.: Paradigms and challenges. In: Greco, S., Figueira, J., Ehrgott, M. (eds.) Multiple Criteria Decision Analysis: State of the Art Surveys, 2 edn., pp. 19–39. Springer, New York (2016)
23. Munier, A.: A Strategy for Using Multicriteria Analysis in Decision-Making. Springer, Netherlands (2011)
24. Kim, S., Park, H., Lee, H., Jung, C.: MAUT approach for selecting a proper decommissioning scenario. In: Proceedings of the Waste Management Symposium 2007 Conference, Tucson, AZ (2007)
25. Tsaples, G., Pruyt, E., Kovari, A., Vasilopoulos, C.: A shock to the system: How can Land Value Taxation change the face of the cities? In: Proceedings of the 31st International Conference of the System Dynamics Society, Boston, MA (2013)
26. Forrester, J.W.: Urban Dynamics. MIT Press, Cambridge, MA (1969)
27. Hobbs, B., Chankong, V., Hamadeh, W., Stakhiv, E.: Does choice of multicriteria method matter? an experiment in water resources planning. Water Resour. Res. **28**(7), 1767–1779 (1992)

# Author Index

Printed in the United States
By Bookmasters